THE SOLAR SYSTEM

THE INNER PLANETS

THE SOLAR SYSTEM
THE INNER PLANETS

SALEM PRESS
A Division of EBSCO Publishing
Ipswich, Massachusetts

GREY HOUSE PUBLISHING

PRINTED IN THE UNITED STATES OF AMERICA

CONTENTS

CONTRIBUTORS

Victor R. Baker

Thomas W. Becker

John L. Berkley

David S. Brumbaugh

Dennis Chamberland

John H. Corbet

John A. Cramer

Anna M. Cruse

Bruce D. Dod

Dave Dooling

Dale C. Ferguson

David G. Fisher

Dennis R. Flentge

John W. Foster

Roberto Garza

Karl Giberson

Gregory A. Good

Alexander A. Gurshtein

Paul A. Heckert

Richard S. Knapp

Joel S. Levine

James C. LoPresto

Randall L. Milstein

Satya Pal

Robert J. Paradowski

Jennifer L. Piatek

George R. Plitnik

Mike D. Reynolds

David M. Schlom

Stephen J. Shulik

R. Baird Shuman

Roger Smith

AURORAS

Category: Earth

Auroras, commonly called the northern and southern lights, are caused by geomagnetic activity taking place in a planet's atmosphere. By understanding auroras, scientists can gauge the effects of solar activities on planetary environments.

Overview

Auroral phenomena were first observed on Earth. Only later were such phenomena detected on other planets in the solar system. "Aurora" is a general term for the light produced by charged particles interacting with the upper reaches of the Earth's atmosphere. The term "aurora borealis" specifically refers to the northern dawn, or northern lights; "aurora australis" refers to the southern lights. Auroras appear in an oval girdling the Earth's geomagnetic poles, where magnetic field lines become nearly perpendicular to the surface. In this region, the Earth is not shielded from the space environment as it is at lower latitudes, where magnetic field lines can be almost parallel to the surface. Thus, electrons and ions moving along magnetic field lines can strike the atmosphere directly. Normally, the auroral oval is located about 23° from the

north magnetic pole and 18° from the south magnetic pole. Because the north magnetic pole is located in Greenland, the oval is offset toward Canada and away from Europe. Generally, auroras appear at altitudes between 100 and 120 kilometers high, in sheets 1 to 10 kilometers thick and several thousand kilometers long.

The auroral oval is a product of the Earth's magnetic field and is driven by the Sun's output of charged particles. The oval can be enlarged as far north or south as 20° latitude; its normal range is around 55-60°. These variations in range and intensity have been correlated with sunspots, showing that solar activity is the engine that drives auroras and other geomagnetic disturbances. Additionally, scientists usually describe auroral activity in terms of local time relative to the Sun rather than the geographic point over which it occurs. Thus, the Earth can be considered to be rotating beneath auroral events (even though the shape of the oval remains skewed). The first indication that auroral displays might be linked to solar activity came in 1859, when Richard Carrington observed an especially powerful solar flare in white light. A few hours later, he observed a strong auroral display and suspected that the two might be linked.

Electrons impinge upon the upper atmosphere from this field-aligned current, moving in a helical path about Earth's magnetic field lines. The helix of electrons trapped in the Earth's magnetic field will become more pronounced as they approach the poles, until finally their direction is reversed at the "mirror point" and they are reflected back to the opposite pole. Motion back and forth is quite normal. If the electrons are accelerated down into the ionosphere, they encounter oxygen atoms and nitrogen molecules. These collisions will release Bremsstrahlung (braking) radiation. These X rays are absorbed by the atmosphere or radiated into space. The oxygen is dissociated from molecular oxygen and then ionized and electrons freed. Those oxygen ions radiate light when neutralized by free electrons. Nitrogen either is excited and radiates when it returns to the "ground" state or is dissociated and excited. As the atmosphere dwindles gradually into the vacuum of space, starting at about 60 kilometers above the surface of the

A view of an aurora, caused by a coronal mass ejection, from the International Space Station. (NASA)

Two curtain-patterned "dueling auroras," as seen over the Yukon in October, 2001. (©Phil Hoffman/Courtesy, NASA)

Earth, the atmosphere forms an electrified layer called the ionosphere, where oxygen and nitrogen molecules are dissociated by sunlight. Because many of these free atoms and molecules are also ionized by sunlight, electric fields and currents move freely, although the net electrical charge is zero.

Auroral structure varies widely. Three major forms have been discerned: quiet or homogeneous arcs, rayed arcs, and diffuse patches. Homogeneous arcs appear as "curtains" or bands across the sky. They sometimes will occur as pairs, and rarely as sets of parallel arcs. They have also been described as resembling ribbons of light. The lower edge of the arc will be sharply defined as it reaches a certain density level in the atmosphere, but the upper edge usually simply fades into space. Pulsating arcs vary in brightness, as energy is pumped in at different rates. Also in the category of quiet arcs are diffuse luminous surfaces, which appear like clouds and have no defined structure; they may also appear as a pulsating surface. Finally, the weakest homogeneous display is a

feeble glow, which actually is the upper level of an auroral display just beyond the horizon.

Auroras with rays appear as shafts of light, usually in bundles. A rayed arc is similar to a homogeneous arc but comprises rays rather than evenly distributed light. The formation and dissipation of individual rays may produce the illusion that rays are moving along the length of a curtain. Among rayed arcs, the "drapery" most resembles a curtain and is most active in shape and color changes. If the viewer is directly below the zenith of an auroral event, then it appears as a corona, with parallel rays appearing to radiate from a central point. Drapery displays are often followed by flaming auroras that move toward the zenith.

A controversial aspect of auroras is whether they produce any sound. Many observers from antiquity to the present have reported hearing auroras; however, sensitive sound-recording equipment has yet to capture such sounds. This leaves open the question of whether or not the sound is a psychological perception, an electrostatic discharge, or some other phenomenon.

Auroral colors—pink, red, green, and blue-green—are distinct and correspond with specific chemistry rather than being a continuous spectrum typical of thermal radiation. Major emissions come from atomic oxygen (at 557.7- and 630-nanometer wavelengths) and molecular nitrogen (at 391.4, 470, 650, and 680 nanometers). These emissions come from distinct altitudes. The green oxygen line (557.7 nanometers), which peaks at 100 kilometers in altitude, is caused by an excited energy state that relaxes in 0.7 second. The red oxygen line (630 nanometers), which peaks at about 300 kilometers, comes from an excited energy state that relaxes in 200 seconds. While oxygen is energized to this level at lower altitudes, its excited energy will be lost to collisions with other gases long before it can relax naturally. From such comparisons, geophysicists have deduced some of the vertical structure of the atmosphere. X rays and ultraviolet light are also emitted but cannot be detected from the ground. The Dynamics Explorer 1 satellite recorded auroras at 130 nanometers in hundreds of images taken several Earth radii above the North Pole. Numerous Earth-sensing satellites continue to observe and study ionospheric physics and auroral phenomena.

Auroral brightness can vary widely. Four levels of international brightness coefficients (IBCs) are assigned, ranging from IBC I, which is comparable to the brightness of the Milky Way, to IBC IV, which equals the illumination received from a full moon. Auroras usually are eighty times brighter in atomic oxygen than in ionized nitrogen molecules, indicating their origins higher in the atmosphere. Doppler shifting is commonly recorded in the spectra around 656.3 nanometers (hydrogen-alpha), indicating the motion of protons that are neutralized and reionized as they accelerate up or down magnetic field lines. It is theorized that only a small fraction (about 0.5 percent) of the energy that goes into auroras actually produces visible light. The remainder goes into radio waves, ultraviolet rays, and X rays, and into heating the upper atmosphere.

Images from the Dynamics Explorer satellite confirmed the indication by ground-based camera chains that auroras are uneven in density and brightness. One image, for example, shows that auroras thin almost to extinction on the dayside but expand to several hundred kilometers in thickness between about 10:00 P.M. and 2:00 A.M. local time. "Theta" auroras have been recorded in which a straight auroral line crosses the oval in the center, giving the appearance of the Greek letter Θ. This phenomenon may be caused by the splitting of the tail of plasma sheets which extends well into the tail of the magnetosphere, or by the solar wind's magnetic field when it has a direction opposite to that of the Earth.

Photographs taken by spin-scan auroral imagers aboard the Dynamics Explorer 1 satellite show that auroral substorms start at midnight, local time, and expand around the oval. Observations of hundreds of substorms show that they have the same generalized structure but that no two are alike. The satellite imager also showed expansions and contractions in the aurora in response to changes in the interplanetary magnetic field and solar wind. As solar wind plasma meets the Earth's magnetosphere, a shock wave is formed, and the wind is diverted around the Earth. This diversion compresses Earth's magnetic field on the sunward side, while it extends like a comet's tail on the nightside. When the field of the solar wind is oriented toward the south, its field lines reconnect with the field lines of the Earth and allow protons and electrons to enter the magnetosphere; they are normally blocked when the field is oriented to the north.

Auroral activity is strongly driven by the solar wind. If the magnetic field of the wind points north—aligned with the Earth's magnetic field—then the auroral oval is relatively small, and its glow is hard to see. When the solar wind's magnetic field reverses direction, a substorm occurs. The oval starts to brighten within an hour, and bright curtains form within it. At its peak, the oval will be thinned toward the noon side and will be quite thick and active on the midnight side. As the storm subsides and

starts to revert to normal about four hours after the field is reversed, the aurora dims and curtains form. Finally, a large, diffuse glow covering the pole may be left as the field becomes stronger in the northward direction.

The flow of the solar wind past the magnetosphere generates massive electrical currents, which flow mostly from one side of the magnetosphere to the other. Some of the currents, however, connect down the Earth's magnetic field, into and through the auroral oval. Because an electric current is caused by the flow of charged particles, in the process electrons are brought directly into the ionosphere around the poles. Primary currents enter around the morning side and exit around the evening side. Secondary currents flow in the opposite direction. Changes in electrical potential of the magnetosphere, as when it is pumped up by particles arriving in the solar wind, will force the electrons through the mirror point. They are then accelerated deeper into the ionosphere. This auroral potential structure is thin but extends around the auroral oval for thousands of kilometers even to the point of closing in on itself.

Electrojets also form in auroras at low altitudes from an effect known as "E-cross-B drift" (written $E \times B$). At high altitudes, electrons and protons flow freely because there is low gas density and no net current change. At lower altitudes, around 100 kilometers, protons are slowed by collisions with atoms and molecules, but electrons continue to move unopposed. The result is a pair of electrojets, eastward (evening) and westward (morning), which flow toward midnight, then cross the polar cap toward noon. These electrojets heat the ionosphere, especially during active solar periods, when auroras are more intense. This $E \times B$ drift in auroral ovals appears to be a major source of plasma for the magnetosphere. It appears that positively charged ions are accelerated upward along the same magnetic field lines, whereas negatively charged electrons precipitate downward. Hydrogen, helium, oxygen, and nitrogen ions compose this flow. Each ion has the same total energy, so their paths vary according to mass. The net effect is that of an ion fountain blowing upward from the auroras which spreads by a wind across the poles.

A little-known subset of the aurora is the sub-auroral red (SAR) arcs, which appear at the midlatitudes; the magnetic field lines on which they occur are different from those on which auroras appear. SAR arcs, which always emit at 660 nanometers (from oxygen atoms), are dim and uncommon. Modern instrumentation has shown that the SAR arcs are a phenomenon separate from the polar auroras. These arcs may be caused by cold electrons in the plasmasphere interacting with plasma waves or with energetic ions. SAR arcs are believed to originate at an altitude of approximately 19,000-26,000 kilometers during especially strong geomagnetic storms, although the arcs themselves appear at altitudes of around 400 kilometers as the energy from the storm leaks or is forced downward.

Auroras also "appear" in the radio spectrum. Studies in the twentieth century showed that auroras could be sounded by radar at certain frequencies. Satellites in the 1970's started recording bursts of energy in the low end of the AM radio spectrum. This radiation is called auroral kilometric radiation (AKR), because its wavelength is up to 3 kilometers, reflected outward by the ionosphere. Bursts can release 100 million to 1,000 million watts at a time, making them far more powerful than conventional broadcasts by humans. Bursts originate in a region of the sky about 6,400-18,000 kilometers high in the evening sector of the auroral oval. Because the radiation is polarized, it is likely that AKR is caused directly by electrons spiraling along magnetic field lines.

Earth is not the only planet to display auroras. Earthly auroral activity has a power rating of approximately 100 billion watts. Auroral displays on Jupiter were detected by the Voyager 1 spacecraft in 1979. During the 1990's the Hubble Space Telescope performed several investigations of this phenomenon. Auroral activity on Jupiter is hundreds of times stronger than on Earth and also appears always to be energized rather than intermittent, as on Earth. In 2007 the Chandra X-Ray Observatory and Hubble Space Telescope conducted a coordinated study of Jovian auroral activity, seeing the phenomenon in both visible and ultraviolet (Hubble) and X-radiation (Chandra) simultaneously. Observing in multiple wavelengths provides clues to the basic physical process involved that seeing an aurora only in visible light cannot.

Saturnian auroras presented a problem with the current understanding of how auroral displays are produced. In 2005 the Hubble Space Telescope and Cassini spacecraft performed coordinated observations of Saturn's auroral activity. Hubble observed in ultraviolet and visible wavelengths, whereas the Cassini probe in orbit about Saturn recorded radio emissions tied to the auroras. An oddity of Saturn's auroras is that, whereas Jupiter's are not affected very much by the solar wind, Saturn's appears to be. Another is that Saturn's auroras brighten on the portion of the planet where darkness leads to sunlight as the storm increases in intensity, which is not the case for Earthly or Jovian auroras.

Uranus's auroras were detected by the Voyager 2 spacecraft's ultraviolet spectrometer. Contemporary studies of Uranian auroras have been performed by the Hubble Space Telescope. The auroral displays of Uranus mimic those on Earth. Uranus's rings have swept clean much of the region that otherwise would have been a rich collection of trapped charged particles that could be taken down along magnetic field lines into Uranus's magnetosphere to generate auroras. Uranus displays both auroras that are centered about its magnetic poles and the sub-auroral red arcs seen on Earth. Indeed, on Uranus the SAR arcs are more prevalent than auroras that are centered about the magnetic poles. The latter variety are believed to result from currents that connect Uranus's rather unusual satellite Miranda to the gas giant's magnetic pole. Uranian auroras generate only weak radio signals.

Voyager 2 detected auroral activity on Neptune. Studies of these displays reveal that Neptune's auroral activity is only half as energetic as that on Earth, despite the disparity in size of the two planets. Also, due to the complexity of Neptune's magnetic field, auroral activity is found on Neptune over areas of the planet far from the magnetic poles.

Ironically, even though the planet has no significant magnetic field, Mars also appears to have auroral displays. Data from Mars Global Surveyor and the European spacecraft Mars Express indicated hundreds of aurora-like displays with less dramatic color variations than those observed on other planets in the solar system. Since a planetary magnetic field is not responsible, some researchers suggest that primordial magnetism associated with patches of the planet's crust, particularly in the southern hemisphere, may be involved in this auroral phenomenon. Martian auroral displays show up mostly in the ultraviolet range, with little or no visible counterpart.

Methods of Study

The space age in some small measure owes its birth to a fascination with auroral phenomena. It was the desire to study and understand the Earth-space interface around the globe at high altitude that resulted in launching the first satellites, during the International Geophysical Year in 1957-1958. Until then, ground-based photography and instrumentation were almost the only methods of studying auroras. Aircraft and rockets played lesser roles. Ground-based instrumentation in the 1940's and 1950's confirmed that auroras were linked to the geomagnetic field, for studies showed that auroras occurred in a circle around the north magnetic pole. Photography of auroral displays has always been difficult because the activity is dynamic, sometimes changing from second to second. Not until the 1950's were electronic devices available to analyze the entire auroral spectrum visible from the ground.

Satellites in the 1970's and 1980's expanded the array of instruments available to investigators. While fields and particles instrumentation has been used to analyze gases and plasmas, imaging instruments have been equally revealing. Notable cameras of various sorts have been carried by Dynamics Explorer 1, the U.S. Air Force HiLat satelliteHiLat (high latitude) satellite, and the Swedish Viking satellite. In addition, some imaging was performed by polar-orbit weather satellites, but with lower spectral and spatial resolutions. The Skylab crews observed some auroral activity. The space shuttle-based Spacelab 3 crew in 1985 photographed auroras from above the atmosphere. Combining photographs taken a few seconds apart allowed formation of stereo imagery so that the structure could be studied better. In other experiments, small

An aurora, caused by both solar plasma and debris from Comet Swift-Tuttle, appeared over the observatory at Mount Megantic in Quebec, Canada, in August, 2000. (Sébastien Gauthier/Courtesy, NASA)

electron guns have been carried into space aboard rockets, on spacecraft, and within the payload bays of space shuttle orbiters. These fire electrons back into the atmosphere in an attempt to generate artificial auroras. An electron gun flew on the first Spacelab (on the space shuttle STS-9 mission) and produced some interesting results. Another such project, on STS-45, was part of the National Aeronautics and Space Administration's (NASA's) INSPIRE program to involve secondary students and undergraduate students in this type of auroral research; unfortunately, although the response of school groups worldwide was strong, the electron gun failed early in the mission.

A key finding from satellite-based research was that auroras are often more active on Earth's dayside, although sunlight and sky completely overwhelm it. Large quantities of radiation are generated in the ultraviolet. This radiation is not seen at the Earth's surface because the atmosphere selectively absorbs such light.

Other planets with magnetospheres also display auroral activity. Such displays are observed almost exclusively by spacecraft either in Earth orbit, like the Hubble Space Telescope, or from spacecraft that either fly by or orbit other planets. The Galileo orbiter routinely picked up auroral activity on Jupiter, and so did the Cassini orbiter on Saturn. Voyager 2 flew past Uranus (1986) and Neptune (1989) and detected auroral activity on these planets as well.

Context

Auroras are the most visible manifestation of the interaction between the Earth and space. The study of plasmas has been enhanced by observations made of them. A clear understanding of auroras will provide a means of diagnosing activities in the magnetosphere and the effects of solar activities on the terrestrial environment.

Auroras also serve as a means to study the magnetohydrodynamics of stars and planets. The physics of planetary auroras are essentially the same as that of Earth's auroral activity, although the energies and chemistries involved may be vastly different. Thus, terrestrial auroras can serve as a laboratory for testing basic theories. Planets with magnetic fields also have auroral activity. Much of Jupiter's radio noise is caused by auroral kilometric radiation, and the Einstein Observatory recorded X rays that apparently came from Bremsstrahlung radiation in the Jovian atmosphere. The Voyager 1 spacecraft observed a 29,000-kilometer-long aurora on the nightside of Jupiter, as well as lightning pulses in and above the clouds coincident with the auroral activity. Voyager 2 did too, and it continued on to the rest of the gas giants in the solar system to do the same. Follow-on spacecraft investigations of these planets, and also on Mars, continued to devote considerable effort to characterizing and understanding the nature of auroral activity production. Each planet's auroral activity provides insight into the nature of that planet's magnetic field and also the plasma environment about it.

Dave Dooling

Further Reading

Akasofu, Syun-Ichi. "The Dynamic Aurora." *Scientific American* 260 (May, 1989). A detailed, college-level treatment of auroras, written by a physicist who is generally accepted as a world expert.

Bone, Neil. *The Aurora: Sun-Earth Interaction*. New York: John Wiley, 1996. One volume in the Ellis Horwood Library of Space Science and Space Technology. Devoted to describing the electrodynamics of the Sun-Earth environment that produce auroral displays.

Bothmer, Volker, and Ioannis A. Daglis. *Space Weather: Physics and Effects*. New York: Springer Praxis, 2006. A selection from Springer Praxis's excellent Environmental Sciences series, this is an overview of the Sun-Earth relationship and provides a historical and technological survey of the subject. Projects the future of space weather research through 2015 and includes information about contemporary spacecraft.

Delobeau, Francis. *The Environment of the Earth*. New York: Springer, 1971. A technical description of the terrestrial environment, written as a reference for space scientists. Although the work is dated by subsequent discoveries, its description of auroral chemistry is still valid.

Dooling, Dave. "Satellite Data Alters View on Earth-Space Environment." *Spaceflight* 29, suppl. no. 1 (July, 1987): 21-29. An article focusing on the exploration of the magnetosphere by the Dynamics Explorer satellites, with details on auroral imaging and radiation.

Moldwin, Mark. *An Introduction to Space Weather*. Cambridge, England: Cambridge University Press, 2008. This text introduces space weather, the influence the Sun has on Earth's space environment, to the nonscientist. Discusses both the scientific aspects of space weather and issues of technological and societal import.

Savage, Candace. *Aurora: The Mysterious Northern Lights*. New York: Firefly Books, 2001. Provides a his-tory of scientific investigation of auroral phenomena. Heavily illustrated with auroral displays.

EARTH'S AGE

Category: Earth

Determining the age of the Earth is one of the great achievements of science. Until the eighteenth century, it was generally believed that the Earth was several thousand years old, and all its features—including mountains and valleys—had been produced by catastrophes such as great floods and earthquakes. The new science of geology gradually showed that the Earth was billions of years old and it had taking its form through slow, uniform processes operating over long periods of time.

Overview

In the middle of the seventeenth century, Joseph Barber Lightfoot of Cambridge University in England penned the following words:

> Heaven and Earth, center and circumference, were made in the same instant of time, and clouds full of water, and man was created by the Trinity on the 26th of October, 4004 B.C.E., at 9 o'clock in the morning.

At the time that Lightfoot wrote those words, this statement expressed the most informed opinion on the age of the Earth. The year 4004 B.C.E. had been calculated by James Ussher, the Anglican archbishop of Armagh, Ireland, by adding up the ages of the patriarchs recorded in the Old Testament. This was the method that most scholars used to date the Earth, and much effort was expended analyzing the first few books of the Old Testament.

A little over a century later, a Scottish physician and gentleman farmer named James Hutton (1726-1797) suggested that there was a better way to determine the past history of the Earth than by poring over biblical genealogies. Hutton believed that processes currently operating in nature could be extrapolated back in time to shed light on the historical development of the Earth. This idea—that past processes are essentially the same as present processes—is called uniformitarianism. In 1785, he presented his new views on geology in a paper entitled "Theory of the Earth: Or, An Investigation of the Laws

Observable in the Composition, Dissolution, and Restoration of Land upon the Globe." Uniformitarianism became the foundation of the newly developing science of geology. Charles Lyell (1797-1875), who was born in the year of Hutton's death, extended these new ideas and helped lay the foundation for what was becoming a powerful new science. A major argument was over the age of the Earth. Was it really millions or billions of years old, as indicated by new discoveries and theories, or was it only a few thousand years old, as everyone had previously believed?

According to current theories, the matter that makes up the Earth as well as everything else in the universe was originally created in the big bang about 13 to 14 billion years ago. Hydrogen, most helium, and trace amounts of lithium and beryllium formed in the immediate aftermath of the big bang, while the atoms of all the other elements were formed by nuclear fusion reactions in massive stars as they generated energy during their "lives" and then exploded as supernovae.

About 4.5 to 4.6 billion years ago, the matter that would become our solar system was part of a nebula, an interstellar cloud of gas and dust, in the disk of the Milky Way galaxy. The portion of this nebula that would become our solar system, called the solar nebula, began to contract under the influence of gravity. Most of the material in the solar nebula collapsed to the center and formed the Sun. The planets and everything else orbiting the Sun formed from the left-over debris through condensation and accretion. All this took place comparatively rapidly, over a time period of a few tens of millions to at most one hundred million years.

Since the initial formation of the Earth, many processes have been taking place in it and on it: Unstable atomic nuclei have radioactively decayed into nuclei of other elements; the Earth's early rotation rate has slowed due to friction from tides generated by the Moon and Sun; mountains have risen under the influence of global tectonics and have been worn away by the ceaseless activities of erosion; and evolution of life has transformed the planet, changing barren wastelands into complex ecosystems teeming with diverse forms of life. These various processes have left their marks on the Earth; by studying them, scientists have begun to reconstruct the history of the Earth, in some cases all the way back to the origin of the Earth about 4.5 to 4.6 billion years ago.

Methods of Study

Many scientific attempts were made during the 1800's to try to determine the age of the Earth. Most involved some process that produced a noticeable change in something. By measuring the magnitude of the change and the rate at which it occurs, the age could be calculated. For example, suppose water is pouring into a bucket at the rate of 1 gallon per minute. If the bucket contains 10 gallons, one can calculate that the process started 10 minutes ago. However, the validity of that result hinges on several assumptions, such as (1) that the rate of water inflow has been constant, (2) that there was no water in the bucket at the start, and (3) that there are no holes in the bucket letting some water drain out.

Similar problems beset the early attempts at dating the Earth. Some of them included estimating how long it would take the Earth to cool and harden from a blob of molten rock and metal, how long it would take to accumulate the entire thickness of sedimentary rocks exposed all around the world, and how long it would take for the oceans to become as salty as they are. Because of uncertainty in the assumptions inherent in these methods, the calculated ages were not at all consistent, ranging from tens of thousands to hundreds of millions of years.

Much more accurate and consistent ages for Earth materials and events have been obtained by radiometric dating, a technique developed during the 1900's. This method uses radioactive decay, a process in which the unstable nucleus of an atom of one element (called the parent) spontaneously transforms into a nucleus of another element (called the daughter). Although it is impossible to say precisely when a single unstable parent nucleus will

Half-Lives of Some Unstable Isotopes Used in Dating

Parent Isotope	Daughter Product	Half-Life Value
Uranium 238	Lead 206	4.5 billion years
Uranium 235	Lead 207	704 million years
Thorium 232	Lead 208	14.0 billion years
Rubidium 87	Strontium 87	48.8 billion years
Potassium 40	Argon 40	1.25 billion years
Samarium 147	Neodymium 143	106 billion years

decay, a large number will decay at a definite rate referred to as half-life. A half-life is the period of time during which half the parent nuclei that are present will decay. Thus after one half-life, one-half of the parent nuclei will have decayed and one half will remain; after two half-lives, three-fourths will have decayed and one-fourth will remain; after three half-lives, seven-eighths will have decayed and one-eighth will remain; and so on. The decay rate or half-life appears to be constant, since nothing seems to have any effect on it. Measuring the ratio of daughter to parent in a sample can tell how long the parent has been decaying and thus how old the specimen is.

The nucleus of an atom is a dense packing of particles called protons with positive electrical charge and neutrons with no electrical charge. All nuclei of a particular chemical element have the same number of protons, but they can have different numbers of neutrons. Nuclei of the same element but with different numbers of neutrons are called isotopes of that element and are identified by their atomic mass number (which is simply the total number of protons and neutrons). For example, all carbon nuclei have six protons. Most carbon nuclei also have six neutrons, and this isotope is called carbon 12. Some carbon nuclei have eight neutrons, and that isotope is called carbon 14.

The three nuclear decay processes underlying geologic radiometric dating are (1) alpha decay, in which the parent nucleus emits an alpha particle, a helium nucleus of two protons and two neutrons; (2) beta decay, in which the parent nucleus emits a beta particle, an electron, when a neutron in the parent nucleus turns into a proton and electron; and (3) electron capture, in which the parent nucleus captures an electron that combines with a proton in the nucleus to form a neutron. The specific isotope decay schemes most commonly used for geologic dating are potassium 40 to argon 40 (half-life 1.3 billion years, via electron capture), rubidium 87 to strontium 87 (half-life 47 billion years, via beta decay), uranium 238 to lead 206 (half-life 4.5 billion years, via eight alpha and six beta decays), uranium 235 to lead 207 (half-life 713 million years, via seven alpha and four beta decays), and thorium 232 to lead 208 (half-life 14.1 billion years, via six alpha and four beta decays). Carbon 14 decays to nitrogen 14 via beta decay with a half-life of only 5,730 years; this is

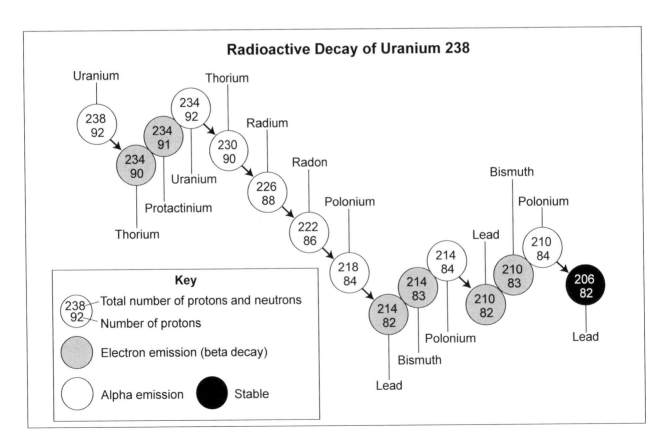

Radioactive Decay of Uranium 238

Key
238/92 — Total number of protons and neutrons
— Number of protons
Electron emission (beta decay)
Alpha emission   Stable

so short that it is limited to dating only very recent geologic events, although it has been very useful for archaeological and historical dating.

Radiometric dating has been applied to thousands of rock specimens from all over the Earth. The oldest rock formation found on Earth so far is the Acasta gneiss from near Great Slave Lake in northern Canada, dated at 4.03 billion years. Even older mineral grains—small crystals of zircon dated at 4.3 to 4.4 billion years—have been found in younger sedimentary rocks from the Jack Hills area of western Australia. It is difficult to find very ancient rocks on the surface of the Earth, because most of the Earth's surface has undergone many changes since the Earth was formed.

The currently accepted age for the Earth, 4.5 to 4.6 billion years, was obtained by radiometrically dating meteorites that fell to Earth from space. These meteorites are believed to be remnants from the time when the planets, including Earth, were forming in the early solar system. Similar dates have been obtained from a few of the rocks brought back by the Apollo landings on the Moon, which is believed to have formed at about the same time as the Earth.

Context

The problem of the age of the Earth is part of a much larger scientific question, which exists at the interface between the study of the Earth and its various processes (which often have practical benefits) and the more esoteric question of the origin and evolution of the universe as a whole. On the practical side, knowledge of the Earth is necessary to predict geologic disasters (such as earthquakes and volcanoes) and to search for geologic resources (such as oil and metallic ores). From a more esoteric point of view, the age of the Earth is important because it speaks to the most fundamental questions that are asked about our place in the universe: How old is this planet, and how was it formed? In the century or two since advances in geological science overthrew the seventeenth century notion of a much younger Earth, people have struggled with finding a new place in the universe. Proponents of "creation science" still argue that the Earth is thousands, not billions, of years old. Legal battles rage over the issue of whether schools across the United States should teach geochronology that is based on religious dogma rather than on scientific research. (In contrast, in Europe this is not a contentious issue, with little public questioning of the antiquity of the Earth based on geologic evidence.) While many questions remain about the details of the formation

of the Earth, two facts seem clear: First, the Earth owes its origin to the same processes that brought the solar system into existence; second, those processes can be dated with a high degree of confidence at between 4.5 and 4.6 billion years ago.

Karl Giberson

Further Reading

Brush, Stephen G. *Nebulous Earth: The Origin of the Solar System and the Core of the Earth from Laplace to Jeffreys*. Cambridge, England: Cambridge University Press, 1997. Describes how thinking about the origin of the solar system changed and includes discussions of the origin of the Earth-Moon system.

Condie, Kent, and Robert Sloan. *Origin and Evolution of Earth: Principles of Historical Geology*. Upper Saddle River, N.J.: Prentice Hall, 1998. An easy-to-read text covering the complexities of the Earth's history, life, and how Earth's subsystems interact. Also discusses dating methods, planetary evolution, and ancient climates.

Dalrymple, G. Brent. *Ancient Earth, Ancient Skies: The Age of the Earth and Its Cosmic Surroundings*. Stanford, Calif.: Stanford University Press, 2004. A book designed for nonscientists who want to learn about the Earth's age. Covers the manner in which scientists collect their data and describes the conclusions to which the data have led them.

Haber, Frances C. *The Age of the World: Moses to Darwin*. Baltimore: Johns Hopkins University Press, 1959. Reprint. Westport, Conn.: Greenwood Press, 1978. Focuses not on estimates of the age of the Earth but rather on the historical controversy that emerged when nonbiblical values for the age of the Earth began to be accepted. Provides insight into the conflict between science and dogma.

Hartmann, William K. *Moons and Planets*. 5th ed. Belmont, Calif.: Thomson Brooks/Cole, 2005. A college textbook beyond the introductory level, with an approach based on comparative planetology. Offers much information on the origin of the solar system in general and the of Earth in particular.

Hurley, Patrick M. *How Old Is the Earth?* Garden City, N.Y.: Doubleday, 1959. One of the few full-length books on geochronology for the layperson. Even though published fifty years ago, it is still useful, as much of the broad overall outline has not changed appreciably since its publication, although many specific details have.

Lewis, Cherry. *The Dating Game: One Man's Search for the Age of the Earth*. Cambridge, England: Cambridge University Press, 2002. The story of Arthur Holmes and the evolution of calculating the age of rocks. Written by a geologist who makes the more technical details understandable to a general audience.

Ozima, Minoru. *The Earth: Its Birth and Growth*. Cambridge, England: Cambridge University Press, 1981. A translation of a Japanese book that was written by a scientist whose specialty is geochronology. Written at an introductory level.

Tarbuck, Edward J., and Frederick K. Lutgens. *Earth: An Introduction to Physical Geology*. Illustrated by Dennis Tasa. 9th ed. Upper Saddle River, N.J.: Pearson Prentice Hall, 2008. This college-level textbook for introductory geology courses is well written and illustrated. It has two chapters on geologic dating and the historical development of the Earth.

Thackray, John. *The Age of the Earth*. New York: Cambridge University Press, 1989. About forty pages long, and published by a British geological museum, this concise volume contains more pictures than text, but the pictures, most in color, are helpful and make this an interesting source.

Wicander, Reed, and James Monroe. *Historical Geology*. 5th ed. Florence, Ky.: Brooks/Cole, 2006. An undergraduate text covering all major areas of historical geology. Provides a history of the Earth and events that have shaped it.

EARTH'S ATMOSPHERE

Category: Earth

The chemical composition of the atmosphere has changed significantly over the history of the Earth. The composition of the atmosphere has been influenced by a number of processes, including interaction with the solar wind; "outgassing" of volatiles (materials that easily vaporize to form gases) originally trapped in the Earth's interior during its formation; the geochemical cycling of carbon, nitrogen, hydrogen, and oxygen compounds between the surface, the ocean, and the atmosphere; and the origin and evolution of life.

Overview

About 4.5 to 4.6 billion years ago, the primordial solar nebula, a part of a large interstellar cloud of gas and dust, began to contract under the influence of gravity. This contraction led to the formation of the Sun and the rest of the solar system including the Earth. The primordial solar nebula was composed mostly of hydrogen gas, with a smaller amount of helium, still smaller amounts of carbon, nitrogen, and oxygen, and still smaller amounts of the rest of the elements of the periodic table.

As the solar nebula contracted, its rotational speed increased to conserve angular momentum. Most of its mass contracted to its center, there becoming the proto-Sun, while the remaining matter was spun off into an equatorial disk. Within the disk, matter condensed from the gaseous state into small, solid grains. Only materials with high melting-point temperatures could condense near the developing proto-Sun; materials with lower melting points condensed farther out. The small solid grains collided with each other and stuck together in a process called accretion that led to the growth of larger bodies called planetesimals. Continued accretion resulted in fewer but larger planetesimals that eventually grew into protoplanets and finally planets, such as Earth.

About the time that the newly formed Earth was reaching its approximate present mass, it may have acquired a temporary atmosphere of hydrogen, helium, methane, ammonia, water vapor, and carbon dioxide—gases that were common in the solar nebula. However, even if such an atmosphere had surrounded the very young Earth, it would have been very short-lived. The Earth's mass was too small to have enough gravity to retain hydrogen and helium for long, and those gases would have quickly escaped into space. Ammonia and methane were chemically unstable in the early Earth's environment and were readily destroyed by ultraviolet radiation from the young Sun. Also, as the young Sun went through its T-Tauri phase of evolution, very strong solar winds (the supersonic flow of protons and electrons from the Sun) would have quickly stripped most of the rest of this primitive atmosphere away.

The early Earth was heated by the intense bombardment of remaining planetesimals and the decay of radioactive elements, to the point where it at least partly melted. The heating and melting released volatiles trapped in its interior by a process called outgassing, forming a gravitationally bound atmosphere. (It is believed that the atmospheres of Mars and Venus also originated in this manner.) The period of extensive volatile outgassing may have

This oblique photograph of Earth, taken from the space shuttle on September 4, 1997, shows the atmosphere and cloud cover over the northwestern African continent. (NASA/JPL/UCSD/JSC)

lasted for many tens of millions of years. The outgassed volatiles probably had roughly the same chemical composition as do present-day volcanic gaseous emissions: by volume about 80 percent water vapor, 10 percent carbon dioxide, 5 percent sulfur dioxide, 1 percent nitrogen, and smaller amounts of hydrogen, carbon monoxide, sulfur, chlorine, and argon.

The water vapor that outgassed from the interior soon reached its saturation point, which is controlled by the atmospheric temperature and pressure. Once the saturation point was reached, the atmosphere could not hold any additional gaseous water vapor. Any new outgassed water vapor that entered the atmosphere would have precipitated out of the atmosphere as rain that fell and formed the Earth's vast oceans. Only small amounts of water vapor remained in the atmosphere—ranging from a fraction of a percentage point to several percent by volume, depending on atmospheric temperature, season, and latitude.

The outgassed atmospheric carbon dioxide, being very water-soluble, readily dissolved into the newly formed oceans and formed carbonic acid. In the oceans, carbonic acid formed ions of hydrogen, bicarbonate, and carbonate. The carbonate ions reacted with ions of calcium and magnesium in the ocean water, forming carbonate rocks, which precipitated out of the ocean and accumulated as seafloor carbonate sediments. Most of the outgassed atmospheric carbon dioxide formed carbonates, leaving only trace amounts of gaseous carbon dioxide in the atmosphere (about 0.035 percent by volume).

Sulfur dioxide, the third most abundant component of volatile outgassing, was chemically transformed into other sulfur compounds and sulfates in the atmosphere. Eventually, the sulfates formed atmospheric aerosols and diffused out of the atmosphere onto the surface.

The fourth most abundant outgassed component, nitrogen, is chemically inert in the atmosphere and thus was not chemically transformed, as was sulfur dioxide. Unlike carbon dioxide, nitrogen is relatively insoluble in water and, unlike water vapor, does not condense out of the atmosphere. For these reasons, nitrogen remained in the atmosphere to become its major constituent (now 78.08 percent by volume). In this way, volatile outgassing led to the formation of the Earth's atmosphere, oceans, and carbonate rocks.

The molecules of nitrogen, carbon dioxide, and water vapor in the early atmosphere were acted upon by solar ultraviolet radiation and atmospheric lightning. In the process, molecules of formaldehyde and hydrogen cyanide could have been chemically synthesized, which would have precipitated and diffused out of the atmosphere into the oceans. In the oceans, the formaldehyde and hydrogen cyanide may have entered into polymerization reactions that eventually led to the chemical synthesis of amino acids, the building blocks of living systems. The synthesis of amino acids from nitrogen, carbon dioxide, and water vapor in the atmosphere and ocean is called chemical evolution. Chemical evolution preceded and provided the material for biological evolution.

There is chemical trace evidence for the existence of microbial living organisms on the Earth by about 3.8 billion years ago; the oldest known simple fossils are at least 3.5 billion years old. These earliest living organisms

were anaerobic since there was no free oxygen in the atmosphere and oceans. Photosynthesis evolved in one or more of these early microbial groups, such as cyanobacteria. In photosynthesis, the organism utilizes water vapor and carbon dioxide in the presence of sunlight and chlorophyll to form carbohydrates, used by the organism for food. In the process of photosynthesis, oxygen is given off as a metabolic by-product. The production of oxygen by photosynthesis was a major event on the Earth and transformed the composition and chemistry of the early atmosphere. As a result of photosynthetic production, oxygen built up to become the second most abundant constituent of the atmosphere (now 20.9 percent by volume).

The evolution of atmospheric oxygen had very important implications for the evolution of life. The presence and buildup of oxygen led to the evolution of respiration, which replaced fermentation as the energy production mechanism in living systems. Accompanying and directly controlled by the buildup of atmospheric oxygen was the origin and evolution of atmospheric ozone, which is chemically formed from oxygen. The production of atmospheric ozone resulted in shielding the Earth's surface from biologically lethal solar ultraviolet wavelengths between about 200 and 300 nanometers. Prior to the evolution of the atmospheric ozone layer, early life was restricted to a depth of at least several meters below the ocean surface. At this depth, the ocean water offered shielding from solar ultraviolet radiation. The development of the atmospheric ozone layer and its consequent shielding of the Earth's surface permitted early life to leave the safety of the oceans and go ashore for the first time in the history of the planet. Theoretical computer calculations indicate that atmospheric ozone provided sufficient shielding from biologically lethal ultraviolet radiation for the colonization of the land once oxygen reached about one-tenth of its present atmospheric level.

Mercury, Venus, and Mars formed in a fashion similar to Earth, but they developed very differently because of their masses and distances from the Sun. They all experienced a period of heating, partial melting, and volatile outgassing of the same gases that led to the formation of the Earth's atmosphere. However, Mercury's distance from the Sun is so close (resulting in high temperatures) that its relatively weak gravity (due to its small mass) was unable to retain more than a thin trace of gases, and thus today it has virtually no atmosphere. In the case of Venus and Mars, the important difference is that the outgassed water vapor never existed in the form of liquid water on the surfaces of those two planets.

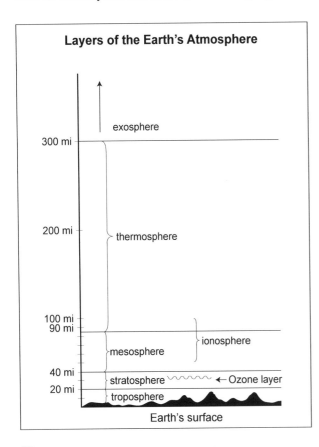

Layers of the Earth's Atmosphere

exosphere

300 mi

thermosphere

200 mi

100 mi
90 mi

ionosphere

mesosphere

40 mi

stratosphere ～～～ ← Ozone layer

20 mi

troposphere

Earth's surface

Storm clouds hover over Earth in this image taken from the International Space Station. (NASA)

Because of Venus's closer distance to the Sun (108 million kilometers versus 150 million kilometers for Earth), its lower atmosphere was too hot to permit the outgassed water vapor to condense out of the atmosphere. Thus, the outgassed water vapor remained in gaseous form in the atmosphere and, over geological time, was broken apart by solar ultraviolet radiation to form hydrogen and oxygen. The very light hydrogen gas quickly escaped from the atmosphere of Venus, and the heavier oxygen combined with surface minerals to form a highly oxidized surface. In the absence of liquid water on the surface of Venus, the outgassed carbon dioxide remained in the atmosphere and built up to become the overwhelming constituent of the atmosphere of Venus (about 96 percent by volume). The outgassed nitrogen accumulated to comprise only about 4 percent by volume of the atmosphere of Venus. This carbon dioxide and nitrogen atmosphere is very massive—it produces an atmospheric pressure at

the surface of the Venus about ninety times the surface pressure of Earth's atmosphere. If the outgassed carbon dioxide in the atmosphere of Earth had not left via dissolution in the oceans and resultant carbonate rock formation, the Earth's surface atmospheric pressure would be about seventy times greater than at present, with carbon dioxide comprising about 98-99 percent of the atmosphere and nitrogen about 1-2 percent. Thus, the atmosphere of Earth would closely resemble that of Venus. The thick carbon dioxide atmosphere of Venus causes a very significant greenhouse temperature enhancement, giving the lower atmosphere and surface of Venus a temperature of about 750 kelvins (about 477° Celsius), which is hot enough to melt lead. For comparison, the average surface temperature of Earth is only about 288 kelvins (about 15° Celsius).

Like Venus, Mars has an atmosphere composed primarily of carbon dioxide (about 95 percent by volume) and nitrogen (about 3 percent by volume). Because of Mars's greater distance from the Sun (228 million kilometers versus 150 million kilometers for Earth), the temperature of the surface of Mars was too low to support the presence of liquid water. There may be very large quantities of outgassed water in the form of ice or permafrost below the surface of Mars. In the absence of liquid water, the outgassed carbon dioxide remained in the atmosphere. The atmospheric pressure at the surface of Mars, however, is only about 7 millibars (the average surface atmospheric pressure on Earth is 1,013 millibars). The smaller mass of the atmosphere of Mars compared to the atmosphere of Venus and Earth may be attributable to the smaller mass of Mars and, therefore, the smaller mass of volatiles trapped in the interior of Mars during its formation. In addition, it appears that the amount of gases trapped in the interiors of Venus, Earth, and Mars during their formation decreased with increasing distance from the Sun. Venus appears to have trapped the greatest amount of gases and was the most volatile-rich planet, Earth trapped the next greatest amount, and Mars trapped the smallest amount.

The atmospheres of the outer planets—Jupiter, Saturn, Uranus, and Neptune—all contain appreciable quantities of hydrogen and helium, along with methane and ammonia. It is believed that the atmospheres of these planets, unlike the atmospheres of the terrestrial planets Venus, Earth, and Mars, are captured remnants of the primordial solar nebula. Because of the outer planets' large masses and their great distance from the Sun resulting in their very low temperatures, hydrogen, helium, methane, and ammonia are stable and long-lived constituents of their atmospheres.

Methods of Study

Information about the origin, early history, and evolution of the Earth's atmosphere comes from a variety of sources. Information on the origin of Earth and other planets is based on theoretical computer simulations. These computer models simulate the collapse of the primordial solar nebula and the formation of the planets. Astronomical observations of what appear to be equatorial disks and the possible formation of planetary systems around young stars have provided new insights into the computer modeling of this phenomenon. Information about the origin, early history, and evolution of the atmosphere is based on theoretical computer models of volatile outgassing and the geochemical cycling and photochemistry of the outgassed volatiles. The process

The SMART 1 spacecraft took this image of Earth from 70,000 kilometers above the surface in May, 2004. Clouds and weather patterns are visible from Scandinavia (top) to northwestern Africa. (European Space Agency)

of chemical evolution—which led to the synthesis of organic molecules of increasing complexity, the precursors of the first living systems on the early Earth—is studied in laboratory experiments. In these experiments, mixtures of gases simulating the Earth's early atmosphere are energized by ultraviolet radiation, electrical discharges, or heated rocks, simulations of energy sources available on the early Earth. The resulting products are analyzed by chemical techniques.

One of the parameters affecting atmospheric photochemical reactions, chemical evolution, and the origin of life was the flux of solar ultraviolet radiation incident on the early Earth. Astronomical measurements of the ultraviolet emissions from young, sunlike stars have provided important information about ultraviolet emissions from the young Sun during the very early history of the atmosphere.

Geological and paleontological studies of the oldest rocks and the earliest fossil records have provided important information on the evolution of the atmosphere and the transition from an oxygen-deficient to an oxygen-sufficient atmosphere. Studies of the biogeochemical cycling of the elements have provided important insights into the later evolution of the atmosphere. Thus, studies of the origin and evolution of the atmosphere are based on a broad cross-section of the sciences, involving astronomy, geology, geochemistry, geophysics, and biology as well as atmospheric chemistry.

Context

Studies of the origin and evolution of the atmosphere have provided new insights into the processes and parameters responsible for global change. Understanding the history of the atmosphere provides insight into its future. Today, atmospheric changes being studied for their possible long-term effects include the buildup of greenhouse gases like carbon dioxide and the depletion of ozone in the stratosphere. The study of the evolution of the atmosphere has provided new insights into the biogeochemical cycling of elements between the atmosphere, biosphere, land, and ocean. Understanding this cycling is a key to understanding environmental problems and possible remedies. Studies of the origin and evolution of the atmosphere have also provided new insights into the origin of life and the possibility of life outside the Earth.

Joel S. Levine

Further Reading

Ahrens, C. Donald. *Essentials of Meteorology: An Invitation to the Atmosphere*. 5th ed. Florence, Ky.: Brooks/Cole, 2007. An updated version of a classic meteorology textbook. Explains tricky concepts in an easy-to-understand way. Suitable for students and nonscientists.

_____. *Meteorology Today*. 8th ed. Florence, Ky.: Brooks/Cole, 2006. A common text used for introductory meteorology college courses but can also be understood by general audiences. Comes with a CD-ROM learning aid, which includes chapter tests and multimedia tutorials.

Chaisson, Eric, and Steve McMillan. *Astronomy Today*. 6th ed. New York: Addison-Wesley, 2008. Very well-written college-level textbook for introductory astronomy courses with an entire chapter on the formation of the planets.

Fraknoi, Andrew, David Morrison, and Sidney Wolff. *Voyages to the Stars and Galaxies*. Belmont, Calif.: Brooks/Cole-Thomson Learning, 2006. A textbook for introductory astronomy courses that offers several sec-

tions dealing with the origin of the solar system and planetary atmospheres.

Freedman, Roger A., and William J. Kaufmann III. *Universe*. 8th ed. New York: W. H. Freeman, 2008. College-level introductory astronomy textbook with several sections on the origin of the solar system, including coverage of planetary atmospheres.

Hartmann, William K. *Moons and Planets*. 5th ed. Belmont, Calif.: Thomson Brooks/Cole, 2005. A college textbook beyond the introductory level, its approach is based on comparative planetology. Includes information throughout the text on the origin of the solar system and its planetary atmospheres.

Henderson-Sellers, A. *The Origin and Evolution of Planetary Atmospheres*. Bristol, England: Adam Hilger, 1983. A technical treatment of the variation of the atmosphere of the Earth over geological time and the processes and parameters that controlled it. Chapters cover the mechanisms for long-term climate change, the atmospheres of the other planets, planetary climatology on shorter timescales, and the stability of planetary environments.

Holland, H. D. *The Chemical Evolution of the Atmosphere and Oceans*. Princeton, N.J.: Princeton University Press, 1984. A comprehensive and technical treatment of the geochemical cycling of elements over geologic time and the coupling between the atmosphere, ocean, and surface. Includes coverage of the origin of the solar system, the release and recycling of volatiles, the chemistry of the early atmosphere and ocean, the acid-base balance of the atmosphere-ocean-crust system, and carbonates and clays.

Levine, Joel S., ed. *The Photochemistry of Atmospheres: Earth, the Other Planets, and Comets*. Orlando, Fla.: Academic Press, 1985. A series of review papers dealing with the origin and evolution of the atmosphere, the origin of life, the atmospheres of Earth and other planets, and climate. The book contrasts the origin, evolution, composition, and chemistry of Earth's atmosphere with the atmospheres of the other planets. It contains two appendixes that summarize all atmospheric photochemical and chemical processes.

Lewis, John S., and Ronald G. Prinn. *Planets and Their Atmospheres: Origin and Evolution*. New York: Academic Press, 1983. A comprehensive treatment of the formation of the planets and their atmospheres. Begins with a detailed account of the origin and evolution of solid planets via coalescence and accretion in the primordial solar nebula; then discusses the surface geology and atmospheric composition of each planet.

Schneider, Stephen E., and Thomas T. Arny. *Pathways to Astronomy*. 2d ed. New York: McGraw-Hill, 2008. A thorough college textbook for introductory astronomy courses, divided into many short sections on specific topics. Contains a unit on the origin of the solar system and several sections on the origins of planetary atmospheres.

Schopf, J. William, ed. *Earth's Earliest Biosphere: Its Origin and Evolution*. Princeton, N.J.: Princeton University Press, 1984. A comprehensive group of papers on such subjects as the early Earth, the oldest rocks, the origin of life, early life, and microfossils. Chapters address the oldest known rock record, prebiotic organic syntheses and the origin of life, Precambrian organic geochemistry, the transition from fermentation to anoxygenic photosynthesis, the development of an aerobic environment, and early microfossils. Technical.

Tarbuck, Edward J., and Frederick K. Lutgens. *Earth: An Introduction to Physical Geology*. Illustrated by Dennis Tasa. 9th ed. Upper Saddle River, N.J.: Pearson Prentice Hall, 2008. This college-level textbook for introductory geology courses is well written and fully illustrated, with a chapter on the origin and historical development of the Earth and its atmosphere.

EARTH'S COMPOSITION

Category: Earth

The Earth consists of a metallic core surrounded by a rocky mantle, which in turn is surrounded by a thin, rocky crust. Much of the crust is covered by an ocean of liquid, salty water. Surrounding it all is the Earth's atmosphere. Of the rocky and metallic material at or below the surface, only the crust, and in a few locales samples of mantle, are available for direct laboratory study. The composition of most of the Earth's interior is inferred by indirect means, primarily from the study of meteorites.

Overview

About 4.5 to 4.6 billion years ago, the solar system was formed from a cloud of gas and dust called the solar nebula. The cloud contracted gravitationally, with most of it forming the Sun. The planets and other objects that today orbit the Sun formed by condensation and accretion in an equatorial disk that developed around the early proto-Sun. Small, solid grains condensed from the gas as it cooled. As the grains in the equatorial disk orbited the proto-Sun, they collided and stuck together, accreting into small bodies called planetesimals. As the planetesimals collided and grew into protoplanets, their gravitational fields increased, so they swept up more material in the equatorial disk. The innermost planets—Mercury, Venus, Earth, and Mars—were formed mainly from dense metals and rocks, while the outer planets—Jupiter, Saturn, Uranus, and Neptune—were formed mostly of gases and volatile ices. During or shortly after Earth's accretion, differentiation occurred; the denser metals, such as iron and nickel, sank to the core of the early Earth, while the less dense rocky material rose to the outer portions of the planet.

Samples of the Earth's crust are readily available. Geologic processes have brought samples from the upper part of the mantle to the Earth's surface in certain locales. Most of the Earth's interior is inaccessible to direct study, but meteorites offer clues to its composition. Most meteorites are remnants of the earliest period of planetary formation. They are classified into three main groups based on composition: stony meteorites, stony-iron meteorites, and iron meteorites. Stony meteorites comprise the most abundant group and are composed of silica-associated, or lithophile, elements such as those found in the Earth's crustal materials. Stony-iron meteorites are composed of roughly equal parts of rock (typically the mineral olivine) suspended in a matrix of iron. Iron meteorites are composed of iron (about 80 to 90 percent) along with siderophile elements such as nickel.

Iron meteorites are particularly suggestive to scientists when they attempt to model the composition of the Earth's core. The average density of the entire Earth is about 5.5 grams per cubic centimeter, while the average density of crustal rocks is only about 2.7 grams per cubic centimeter for continental crust and 3.0 grams per cubic centimeter for oceanic crust. This simple comparison indicates that the core must be substantially denser than the average for the entire Earth, and the only reasonably abundant element with about the right density is iron.

The core has two parts: a solid inner core, with a radius of 1,300 kilometers and a density of about 12 to 13 grams per cubic centimeter, and a molten outer core, 2,200 kilometers thick, with a density of about 10 grams per cubic centimeter. The inner core is mostly iron and nickel under high pressure (to make it solid), while the molten outer

Chemical Composition of Earth's Crust

Element	Weight (%)	Volume (%)
Oxygen (O)	46.59	94.24
Silicon (Si)	27.72	0.51
Aluminum (Al)	8.13	0.44
Iron (Fe)	5.01	0.37
Calcium (Ca)	3.63	1.04
Sodium (Na)	2.85	1.21
Potassium (K)	2.60	1.88
Magnesium (Mg)	2.09	0.28
Titanium (Ti)	0.62	0.03
Hydrogen (H)	0.14	—

Separated into three main layers—the core, mantle, and crust—the Earth is an active body, its internal heat far from exhausted. The complexity of the chemical composition increases with each successive outward layer. This generalized model gives a framework for examining the relationships of Earth materials.

Earth's wide range of pressure and temperature regimes helps explain why several thousand distinct minerals and numerous rock types composed of different combinations of minerals have been recognized in samples of the crust and upper mantle. Sampling a variety of crustal rocks leads to a determination of elemental abundance in the crust. By mass, approximately one-half of the crust is oxygen and approximately one-fourth is silicon. These two elements, plus aluminum, iron, calcium, sodium, potassium, and magnesium, make up more than 99 percent of the Earth's crust. Silicon and oxygen combine to form the silicon-oxygen tetrahedron, consisting of a single silicon atom surrounded by four oxygen atoms evenly spaced around it three-dimensionally at the corners of a tetrahedron. This silicon-oxygen tetrahedron joined to additional tetrahedra and/or atoms of other elements forms the class of minerals called silicates, by far and away the most common minerals in the crust.

core probably contains, besides iron and nickel, lighter elements such as sulfur, silicon, oxygen, carbon, and hydrogen. As a whole, the core comprises about one-sixth of the Earth's volume and about one-third of the Earth's mass.

Almost all the remaining two-thirds of the Earth's mass is contained in the mantle, making the mass of the crust, oceans, and atmosphere insignificant in comparison. The mantle is rich in dense, ultramafic rocks such as peridotite, composed mostly of the minerals olivine and pyroxene.

Shortly after (or perhaps during) the initial condensation of grains and their accretion into planetesimals and protoplanets, the Earth's thermal history began through the process of radioactive decay. During this early thermal period, radioactive nuclides (atoms of specific isotopes) decayed, producing substantial heat that led to at least partial melting. Much of the heating is attributable to the decay of potassium 40 and short half-life elements such as aluminum 26. After as little as perhaps 100,000 years, the planet separated into the iron-nickel core and magnesium-iron-silicate lower mantle. Over a longer timescale (probably more than ten million years but no more than a few hundred million), the high-volatility compounds (such as lead, mercury, thallium, bismuth, water in hydrated silicates, carbon-based organic compounds, and the noble gases) all migrated to the surface, where the material was outgassed or melted into magmas in a continuous period of crustal reprocessing that lasted for several hundred million years.

As ultramafic magmas cool, successive minerals crystallize and settle out via reaction series. As the temperature drops in the melt zone, a discontinuous series (a set of discrete reactions) can occur. Magnetite, an oxide of iron and titanium, is the first to settle out, at about 1,400° Celsius (1,700 kelvins). Olivine, a silicate mineral with a crystal lattice structure of individual silicon-oxygen tetrahedra joined together by other ions (commonly iron and magnesium) and a density between 3.2 and 4.4 grams per cubic centimeter, is the next to crystallize out of the melt. Then comes pyroxene, a silicate mineral with its silicon-oxygen tetrahedra connected in long single chains and a density of 3.2 to 3.6 grams per cubic centimeter. As temperatures in the magma drop to near 1,000° Celsius (1,300 kelvins), the next to crystallize is amphibole, a silicate mineral with its silicon-oxygen tetrahedra joined in long double chains and a still lower density of 2.9 to 3.2 grams per cubic centimeter. As the cooling progresses, the lattice structures increase in complexity with biotite mica, with its silicon-oxygen tetrahedra joined in planar sheets.

Paralleling this discontinuous series of reactions is the continuous reaction series of plagioclase feldspar. It has a full three-dimensional lattice of silicon-oxygen tetrahedra, and it varies continuously from being calcium-rich at high temperatures of crystallization to sodium-rich at lower temperatures. Finally, at still lower temperatures down to about 1,000 kelvins (700° Celsius), come potassium feldspar, muscovite mica, and quartz.

With this information, one can start to hypothesize about how the crust and its ocean basins and continents evolved. The oldest Earth materials yet identified are zircon crystals, possibly dating back 4.4 billion years, found in the Jack Hills area of Australia, while the oldest known continental rocks—the Acasta gneiss from the Northwest Territories of Canada—are about 4 billion years old, and they were metamorphosed from earlier igneous rocks. This means that within a few hundred million years after the initial formation of the Earth through condensation, accretion, and differentiation, the first crustal rocks of the Archean eon formed. They probably were composed of olivine, pyroxene, and anorthite (calcium-rich plagioclase feldspar), which crystallized out of basaltic magmas that rose to the surface and cooled and hardened. The early crust, which may have been similar to the anorthosite that makes up much of the ancient highlands on Earth's moon, formed a sheet that was fractured into pieces and subjected to heating through radioactive decay. Differentiation led to the formation of thicker granitic regions surrounded by the thinner basaltic crust. This was the beginning stage in the development of today's crust, which consists of two main types: the denser, thinner, mafic or basaltic oceanic crust and the less dense, thicker, felsic or granitic continental crust. The onset of plate tectonics moved the early continental fragments, causing them to collide and weld themselves together into continental shields in episodes of mountain-building, called orogenies.

The Earth's original inventory of gases appears to have been lost very early in its history, to be replaced with a secondary atmosphere through volcanic outgassing and perhaps impacts of volatile-rich cometary nuclei and carbonaceous chrondrite meteorites. Extensive volcanic activity and high surface temperatures gradually diminished until the hydrosphere (water cycle) was established and oceans appeared.

Life on Earth existed at least 3.5 billion years ago, as evidenced by microfossils similar to modern cyanobacteria (blue-green algae). With the oceans growing in volume and salinity and the development of oxygen-releasing life-forms, Earth's geochemistry became more complex. By the beginning of the Paleozoic era, about 540 million years ago, the oxygen content of

Primary Rocks and Minerals in Earth's Crust

Rocks	% Volume of Crust	Minerals	% Volume of Crust
Sedimentary		Quartz	12
Sands	1.7	Alkali feldspar	12
Clays and shales	4.2	Plagioclase	39
Carbonates (including salt-bearing deposits)	2.0	Micas	5
		Amphiboles	5
		Pyroxenes	11
Igneous		Olivines	3
Granites	10.4	Clay minerals (and chlorites)	4.6
Granodiorites, diorites	11.2		
Syenites	0.4	Calcite (and aragonite)	1.5
Basalts, gabbros, amphibolites, eclogites	42.5	Dolomite	0.5
		Magnetite (and titanomagnetite)	1.5
Dunites, peridotites	0.2	Others (garnets, kyanite, andalusite, sillimanite, apatite, etc.)	4.9
Metamorphic			
Gneisses	21.4		
Schists	5.1		
Marbles	0.9	**Totals**	
		Quartz and feldspars	63
Totals		Pyroxene and olivine	14
Sedimentary	7.9	Hydrated silicates	14.6
Igneous	64.7	Carbonates	2.0
Metamorphic	27.4	Others	6.4

Source: Michael *H.* Carr et al., The Geology of the Terrestrial Planets, NASASP-469, 1984. Data are from A. B. Ronov and A. A. Yaroshevsky, "Chemical Composition of the Earth's Crust," American Geophysical Union Monograph 13.

the atmosphere had reached 1 percent of its present level. Life-forms significantly shaped the Earth's chemical composition. Multicelled animals in the oceans scrubbed carbon dioxide from the atmosphere and locked it up in the carbonate rocks, forming biochemically precipitated limestones. By the latter part of Paleozoic era, about 300 million years ago, coal formed as a result of the first land forests being periodically inundated by ocean transgressions.

Methods of Study

Perhaps no other Earth science is as speculative as that of early Earth history and the geochemical evolution of the Earth. Some of the major challenges confronting

Typical Composition of Rocks That Compose Much of the Earth's Mantle or Crust

Oxide	Unmelted Peridotite in the Mantle	Basalt Formed at Oceanic Ridges or Rises	Andesite Formed at Subduction Zones	Granite Rock Along Continental Subduction Zones	Continental Rift Basalt	Shale	Sandstone Near the Source	Sandstone Far from the Source	Limestone
SiO2 (silicon oxide)	45.0	49.0	59.0	65.0	50.0	58.0	67.0	95.0	5.0
TiO2 (titanium oxide)	0.4	1.8	0.7	0.6	3.0	0.7	0.6	0.2	0.1
Al2O3 (aluminum oxide)	8.7	15.0	17.0	16.0	14.0	16.0	14.0	1.0	0.8
Fe2O3 (ferric iron oxide)	1.4	2.4	3.0	1.3	2.0	4.0	1.5	0.4	0.2
FeO (ferrous iron oxide)	7.5	8.0	3.3	3.0	11.0	2.5	3.5	0.2	0.3
MnO (manganese oxide)	0.15	0.15	0.13	0.1	0.2	0.1	0.1	—	0.05
MgO (magnesium oxide)	28.0	8.0	3.5	2.0	6.0	2.5	2.0	0.1	8.0
CaO (calcium oxide)	7.0	11.0	6.4	4.0	9.0	3.0	2.5	1.5	43.0
Na2O (sodium oxide)	0.8	2.6	3.7	3.5	2.8	1.0	2.9	0.1	0.05
K2O (potassium oxide)	0.04	0.2	1.9	2.3	1.0	3.5	2.0	0.2	0.3
Volatiles (water or carbon dioxide)	1.0	1.0	1.0	2.0	1.0	8.0	2.0	1.0	42.0

Note: Compositions are given as weight percentages of the element oxide in the entire rock.

Earth scientists are questions about how the Earth's crust formed and when plate tectonic movement began. It is generally accepted by most Earth scientists that heat flow was substantially greater and hence crustal formation occurred more rapidly in Archean times. Despite the problems of extrapolating back to a time when the first solid rocks were forming, the established models are based on some solid lines of evidence.

In 1873, American geologist James D. Dana made one of the initial advances in the study of the Earth's internal chemical composition when he suggested that analogies could be drawn from the study of meteorites. Geochemists studying meteorites today have derived radiometric dates of 4.4 to 4.6 billion years for many of them—corresponding to the initial epoch of condensation and accretion in the solar nebula. Because meteorite types approximate the elemental distribution in the Earth, they are valuable samples of what the Earth formed from.

Geophysicists use seismic waves from earthquakes to study the structure of the Earth's interior. Variations in speed as the waves pass through the Earth, and reflection and refraction of them at internal boundaries, have revealed a differentiated Earth with a very dense metallic core, a less dense rocky mantle, and an even less dense rocky crust "floating" on top. The well-established theory of plate tectonics holds that the crust and upper mantle together form rigid lithospheric plates that are moving, driven by slow convection currents in the mantle.

The drive to study Archean rocks was partly fueled by the United States Apollo missions to the Moon, which returned rocks of comparable age from the lunar surface. Interest in Archean crustal evolution was further aroused by the discovery of Archean lavas called komatiites around greenstone belts (which are agglomerations of Archean basaltic, andesitic, and rhyolitic volcanics, along with their sediments derived by weathering and erosion). Komatiites are ultramafic lavas that formed at temperatures greater than about 1,100° Celsius (1,400 kelvins) and may be fragments of the first crust. Work by field geologists in regions with exposed Archean rocks found successively older granitic rocks—3.8 billion years in western Greenland, 3.9 billion years in Antarctica, and 4.0 billion years in Canada's Northwest Territories. Even older detrital zircons with radiometric ages between 3.8 and 4.4 billion years were discovered in somewhat younger sedimentary rocks in western Australia. The zircon find is significant because it sets an approximate birth date for early continental crust, as zircon is a reasonably common though minor constituent of granitic

igneous rocks. The Australian zircons probably formed in early continental igneous rocks and then were eroded, transported, and deposited with other sediments in the sandstones in which they were found.

Geochemists have refined their study of these ancient rocks with more sophisticated methods to determine isotope ratios in them. Instruments common in geochemical laboratories today use X-ray diffraction and gamma-ray spectral analysis to determine which isotopes are present. Isotope ratios in rocks are of particular interest to geochemists because they provide clues as to chemical cycles in nature. The equilibria of these cycles, as indicated by the isotope ratios, offer insights into volcanic, oceanic, biological, and atmospheric cycles and conditions in the past.

Context

Perhaps no other area of scientific study is as intriguing and controversial as that of the origin and evolution of the Earth. Geochemists and geophysicists have been at the forefront of the quest to understand the Earth's present geology in terms of its past. Before the 1960's, little was known of the Earth's history during early Precambrian times. This lack is significant when one considers that the Precambrian comprises about eight-ninths of the geologic timescale.

It is likely that improved techniques used to analyze rocks and minerals in the laboratory will continue to provide a better understanding of the formation of the Earth's crustal materials and the evolution of moving lithospheric plates. Radiometric dating and isotope analysis will help unravel the relationships between the greenstone belts and granulite-gneiss associations that typify Archean formations on all continents.

Studying features and materials on other solar system bodies will also lead to a better understanding of the early Earth and its evolution. Similarities and differences in Earth's early history are expected to be revealed by future space probes to the Moon, Mars, Venus, Mercury, and asteroids.

David M. Schlom

Further Reading

Fyfe, W. S. *Geochemistry*. Oxford, England: Clarendon Press, 1974. Part of the Oxford Chemistry series, this work was written for lower-division college chemistry students. Although in some respects dated, it is nevertheless a brief (about one-hundred-page) and excellent introduction to the science of geochemistry. Of special interest is chapter 9, "Evolution of the Earth." Bibliography, glossary, index.

Gregor, C. Bryan, et al. *Chemical Cycles in the Evolution of the Earth*. New York: John Wiley & Sons, 1988. A systems approach to geochemistry, this book is suitable for the serious college student. Although filled with graphs, tables, and chemical equations, sections are still accessible to the layperson as well. Discussions of mineralogical, oceanic, atmospheric, and other important chemical cycles are extensive and the work is well referenced.

Hartmann, William K. *Moons and Planets*. 5th ed. Belmont, Calif.: Thomson Brooks/Cole, 2005. A college textbook beyond the introductory level, its approach is based on comparative planetology. Provides much information on condensation and accretion in the solar nebula, as well as the composition of the terrestrial planets generally and Earth in particular.

Kroner, A., G. N. Hanson, and A. M. Goodwin, eds. *Archaean Geochemistry: The Origin and Evolution of the Archaean Continental Crust*. Berlin: Springer, 1984. A collection of reports by the world's leading geochemists studying the geochemistry of the world's oldest rocks. Although many of the articles are technical in nature, the abstracts, introductions, and summaries are accessible to a college-level reader interested in the work of top international scientists.

Levin, Harold L. *The Earth Through Time*. 5th ed. Fort Worth: Saunders College Publishing, 1996. A thorough and readable college text on historical geology. Filled with illustrations, photographs, and figures, this book is also suitable for the layperson. Chapters on planetary beginnings, origin and evolution of the early Earth, and plate tectonics are of special interest. Contains an excellent glossary and index.

Salop, Lazarus J. *Geological Evolution of the Earth During the Precambrian*. New York: Springer-Verlag, 1983. A top Soviet geologist conducts an exhaustive survey of Precambrian geology. Suitable for a college-level reader with a serious interest in the subject. Contains numerous graphs and tables, with extensive references.

Tarbuck, Edward J., and Frederick K. Lutgens. *Earth: An Introduction to Physical Geology*. Illustrated by Dennis Tasa. 9th ed. Upper Saddle River, N.J.: Pearson Prentice Hall, 2008. This college-level textbook for introductory geology courses is well written and illustrated. It has a full chapter on the origin, historical development, and composition of the Earth.

Wedepohl, Karl H. *Geochemistry*. New York: Holt, Rinehart and Winston, 1971. An older but still good and accessible brief introduction to geochemistry fundamentals. Contains an excellent chapter on meteorites and cosmic abundances of the elements. Suitable for the nontechnical reader, with index and references. A good starting point for those unfamiliar with mineral formation.

EARTH'S MAGNETIC FIELD AT PRESENT

Category: Earth

The study of the Earth's magnetic field is important from both academic and practical perspectives. The Earth is the only planet of the inner solar system with a strong magnetic field, and this provides clues about the formation of the Earth and the other inner planets. The Earth's magnetic field deflects and traps high-energy charged particles, providing a shield to protect life. It can disrupt modern communication and electrical systems, but it also can point to the location of ore deposits.

Overview

The study of the Earth's magnetic field is a branch of geophysics, which combines geology and physics to investigate various physical characteristics of the Earth. The ultimate source of any magnetic field is moving electrical charge, such as an electric current flowing in a wire. Approximately 90 to 95 percent of the Earth's magnetic field is thought to be produced by electrical currents in the Earth's molten metallic outer core, a mechanism referred to as the geodynamo.

The Earth's field is predominantly a dipole field, meaning it has two magnetic poles; the prefix "di" is derived from the Greek word meaning "two." This is the type of field produced by a bar magnet or an electric current flowing in a wire loop. By definition, the north pole of a bar magnet is the end that points northward on Earth at the present time. Since like magnetic poles repel and unlike magnetic poles attract, that means the Earth's magnetic south pole is located in the Northern Hemisphere, and the Earth's magnetic north pole is located in the Southern Hemisphere. Magnetic field lines are a way of visualizing the direction of a magnetic field, and by convention they point in the direction the north pole of a bar magnet would point. Magnetic field lines leave the Earth's surface in the Southern Hemisphere, arc over the Earth, and reenter the Earth in the Northern Hemisphere. The magnetic poles are the two places where the field lines leave and enter the Earth's surface precisely vertically.

The pole in the Northern Hemisphere is called the north magnetic pole (but remember it is a magnetic south pole), and the pole in the Southern Hemisphere is called the south magnetic pole (although it is a magnetic north pole). At first this may seem confusing, but note the distinction in terminology: the geographic hemisphere that the pole is in comes before the words "Magnetic Pole," while the type of pole comes between words "magnetic" and "pole." The field's strength is about 0.6 gauss (a unit of magnetic induction) at the magnetic poles and about 0.3 gauss at the magnetic equator, where the field lines are horizontal. (For comparison, a small bar magnet has a field strength of about 1 gauss.) The difference in strength is due to the field lines bunching together at the magnetic poles and spreading apart at the magnetic equator.

The magnetic poles are not located at the geographic or rotational poles of the Earth, which are the two points where the rotational axis of the Earth intersects the surface. Currently, near the beginning of the twenty-first century, the north magnetic pole is located in the Arctic Ocean north of Canada and west of Greenland, approximately 900 kilometers from the geographic North Pole. The south magnetic pole is located in the ocean between Antarctica and Australia, approximately 2,900 kilometers from the geographic South Pole. Notice that the magnetic poles are not symmetrically located in relation to the rotational poles.

The magnetic poles also are not stationary; rather, they wander around the polar regions at varying speeds of up to tens of kilometers per year. Since the early 1900's, the north magnetic pole has moved roughly north-northwest about 1,300 kilometers, while the south magnetic pole has moved from Antarctica northward out into the ocean toward Australia. Because the geodynamo, the theoretical source of the Earth's magnetic field, is driven partly by the Earth's rotation, it is presumed that over long periods of time, the positions of the two magnetic poles average out to roughly the locations of the geographic poles. In addition, measurements of the field's strength since the mid-nineteenth century indicate that it is decreasing at a rate of about 6 percent per century. Archaeomagnetic evidence indicates the field was approximately twice as strong two millennia ago, and before that, around 3,500 B.C.E., it was only about one-half the present strength.

These changes in direction and strength over timescales of years to millennia are called secular variations. They are thought to be due to changes in the geodynamo operating in the Earth's molten outer core. Considering its past behavior, scientists cannot predict what the magnetic field will do in the future. It may continue to decrease, or it may increase. If it were to continue decreasing at the present rate, the field would drop to zero in about 1,600 years. This might lead to a magnetic reversal, in which the field re-forms but with its polarity reversed. Paleomagnetic measurements of past magnetic fields preserved in some rocks indicate this has occurred many times in the geologic past, the last time about 700,000 years ago.

The Earth's magnetic field also exhibits small, rapid changes in direction and strength over periods of hours to days, due to a variety of external effects. For example, the gravitational fields of the Sun and Moon distort the atmosphere of the Earth, in the same manner as ocean tides. Movement of electrically charged particles in the atmosphere produces a weak contribution to the magnetic field that changes with the relative positions of the Sun and Moon.

The Sun continually blows electrons, protons, and other electrically charged particles outward from its surface at speeds of hundreds of kilometers per second, a phenomenon known as the solar wind. When these charged particles encounter the Earth's magnetic field, they interact with it, producing a boundary called the magnetopause. Inside the magnetopause is the magnetosphere, the region in which the Earth's magnetic field is dominant. The solar wind changes the shape of the Earth's field. The side facing the Sun is pushed in toward the Earth by the solar wind so that the magnetopause is about 60,000 kilometers, or 10 Earth radii, from the Earth, while the field pointing away from the Sun is elongated into a magnetic tail that can extend farther than the orbit of the Moon.

Some of the solar wind particles, particularly electrons and protons, are trapped by the Earth's magnetic field. These form the Van Allen belts, which were discovered in 1958 by Dr. James Van Allen while analyzing data from a charged particle detector he had placed aboard Explorer 1, the first successful U.S. satellite. The inner belt is a torus about 3,000 kilometers above the magnetic equator; the outer belt is a larger torus about 14,000 kilometers above the magnetic equator.

The number of sunspots increases and decreases over a cycle of eleven years. Sunspots are just one of the more obvious manifestations of solar magnetic activity, and the Sun reverses magnetic polarity with each eleven-year cycle. During times of maximum solar activity, solar flares are most likely to erupt from its surface. These flares eject large numbers of highly energetic, electrically charged particles out into the solar system. If they encounter the Earth's magnetic field, they can produce magnetic storms that cause wild variations in the Earth's field. This in turn can disrupt modern communication and electrical distribution networks. It is at these times, when the Sun is most active, that auroras (the northern and southern lights) are most common. Increased numbers of charged particles from the Sun are deflected by the Earth's magnetic field and enter the Earth's upper atmosphere near the magnetic poles, where they excite air molecules, causing them to glow.

Lightning is a very rapid electrical discharge in the atmosphere; electrical charges can flow from the ground to clouds, from clouds to the ground, or from cloud to cloud. Locally, this strong but brief electrical current produces a very large increase and then decrease in the background field strength.

An artist's conception of the outflowing solar wind meeting Earth's magnetic field. (NASA/ESA)

Magnetic anomalies distort the dipole shape of the main field. Some of these anomalies probably result from more complicated flow patterns in the molten outer core, while others probably are associated with rock units that are rich in iron. Two of the strongest known are located near Kursk, Russia, and in northern Manitoba, Canada. Running parallel to the ocean ridge-rift system are bands or strips of seafloor with alternate normal and reversed magnetic polarity that enhance or weaken the present field over them. The strips preserve a record of the Earth's past magnetic field, frozen into the igneous rocks (mainly basalt and gabbro) that cooled from lava that oozed out along the ridge-rift, and then was pushed away from the ridge-rift as new lava oozed out. This provides evidence of magnetic field reversals in the geologic past and support for the concept of seafloor spreading (one of the key parts of plate tectonics). Small anomalies can even result from man-made iron objects.

Methods of Study

The orientation of the magnetic field at any point on Earth is specified by two angles called declination and inclination. Declination is the angle between true north (the direction of the geographic or rotational north pole) and the horizontal component of the magnetic field line at that point. Thus declination is the angle between true north and the direction an ordinary compass needle points. Inclination is the angle between a horizontal line and the downward tilt of the magnetic field line at that point. Inclinations are downward (positive) in the northern hemisphere and upward (negative) in the Southern Hemisphere. The magnetic poles are located where the inclination is 90°, specifically 90° down (positive) for the pole in the Northern Hemisphere and 90° up (negative) for the pole in the Southern Hemisphere. The magnetic equator is located where the inclination is 0°.

Around the world, 130 permanent magnetic observatories have been established to record any changes in the magnetic field. It was at observatories in London and Paris that secular variations of the field were first recognized in the 1600's. Early observatories could measure only the declination and inclination of the field. Declination was measured with a compasslike device and inclination with a magnetized rod balanced so that it could pivot freely in a vertical plane.

Magnetometers for the measurement of magnetic field intensity were first developed in the mid-1800's, and a number of different types are in use today. In conjunction with the magnetic observatories on the ground, some satellites carry magnetometers for the measurement of the field from orbit, and they provide readings for virtually the entire globe.

Portable magnetometers can detect local field anomalies due to things under the surface. Geologists use them to prospect for magnetic iron ore deposits, and archaeologists use them to search for buried iron artifacts.

Context

When magnetic storms occur, modern communication and electrical distribution networks can be disrupted. Also on these occasions, auroras are more likely to occur and be seen over larger areas. The magnetic field interacts with electrically charged particles and prevents many of them from reaching the Earth's surface. It is possible that a decrease in the field would lead to more particles reaching the surface, perhaps producing greater numbers of genetic mutations or cancers. Changes in the field strength have been suggested as a cause of some of the mass extinctions that have occurred in the geologic past.

Stephen J. Shulik

Further Reading

Courtillot, V., and J. L. Le Mouel. "Time Variations of the Earth's Magnetic Field: From Daily to Secular." In *Annual Review of Earth and Planetary Sciences* 16 (May, 1988): 389-486. This source covers the geomagnetic field in general and then provides an in-depth study of its variations, from short-term to very long-term. Very little mathematics; many figures.

Fowler, C. M. R. *The Solid Earth: An Introduction to Global Geophysics*. 2d ed. New York: Cambridge University Press, 2004. An updated version of a widely used textbook for introductory geophysics courses. Designed for students with some knowledge of physics and calculus.

Garland, G. D. *Introduction to Geophysics*. 2d ed. London: W. B. Saunders, 1979. Used as a text for introductory geophysics, this book contains in chapter 17, "The Main Field," readable material on the main field and its generation. Time variations are discussed

as well as the external field and methods of measurement. Some equations; many figures and graphs of interest to the less specialized reader.

Jacobs, J. A., R. D. Russell, and J. T. Wilson. *Physics and Geology*. 2d ed. New York: McGraw-Hill, 1974. This introductory geophysics textbook is formidable for the average student because it uses considerable mathematics in some chapters, but chapter 8, "Geomagnetism," has sections on the present field and contains a minimum of equations and many figures and graphs. Auroras and the magnetosphere are also discussed.

Knecht, D. J., and B. M. Shuman. "The Geomagnetic Field." In *Handbook of Geophysics and Space Environment*, edited by A. S. Jursa. Springfield, Va.: National Technical Information Service, 1985. This source covers the geomagnetic field and various aspects of it: terminology, sources of the field, measurements, the main field, and sources of geomagnetic data. Some sections have no mathematics, but others have a small amount. Many figures help the reader to understand the authors' narratives. A number of references are listed at the end of the chapter.

Motz, Lloyd, ed. *Rediscovery of the Earth*. New York: Van Nostrand Reinhold, 1979. As a collection of articles for the nonscientist by scientists renowned in their respective fields, the text makes interesting reading, augmented with many colorful illustrations. The chapter "The Earth's Magnetic Field and Its Variations" is written by Dr. Takesi Nagata, who has written hundreds of articles on diverse aspects of geophysics besides the Earth's magnetic field.

Smith, David G., ed. *The Cambridge Encyclopedia of Earth Sciences*. New York: Cambridge University Press, 1982. Chapter 7, "The Earth as a Magnet," contains information about the Earth's present-day magnetic field, geomagnetic field changes, and magnetic anomalies. The text is well written at a nontechnical level, with many colorful diagrams and figures.

Stacey, F. D. *Physics of the Earth*. New York: John Wiley & Sons, 1977. In section 8.1, "The Main Field," the author provides a short, technical description of the main field that is of interest to the more advanced student. As a textbook for geophysics, it covers many other areas on a technical level.

Tarbuck, Edward J., and Frederick K. Lutgens. *Earth: An Introduction to Physical Geology*. Illustrated by Dennis Tasa. 9th ed. Upper Saddle River, N.J.: Pearson Prentice Hall, 2008. This college-level textbook for introductory geology courses is well written and illustrated. The chapter on the Earth's interior has a section on the Earth's magnetic field.

Vogel, Shawna. *Naked Earth: The New Geophysics*. New York: Plume, 1996. Covers geophysics research and theories since 1960. Includes information about Pangaea, the supercontinent cycle, and the reversals of the Earth's magnetic field. For general audiences.

EARTH'S MAGNETIC FIELD: ORIGINS

Category: Earth

The Earth has a dipole magnetic field that is roughly aligned with its rotational axis. A dynamo effect in the Earth's molten outer core is the most likely source of most of the magnetic field.

Overview

The Earth's magnetic field is primarily a dipole field, meaning it has two well-defined magnetic poles, called "north" and "south" (the prefix "di" comes from a Greek term for "two"). This is the type of field produced by a bar magnet or an electric current flowing in a wire loop. The Earth's iron core once was thought to act like a giant bar magnet, possibly due to a remanent field frozen in place from some primordial magnetic field that existed when our solar system was forming. Now, however, electrical currents produced by fluid motions in the molten outer part of the Earth's iron core are theorized to be the source of the magnetic field. This conclusion is based on models of the Earth's interior structure and variations in the magnetic field over both historic and geologic timescales.

The ultimate source of any magnetic field is the movement of electric charges. Wires carrying electric currents, for example, have magnetic fields around them because of the electric charges (electrons) moving through the wires. The electrons surrounding the nucleus of an atom are moving, and this produces a minute magnetic field. In a magnet, the atoms are aligned in such a way that these small fields add together to produce the larger field of the magnet. The conclusion, therefore, is that electric currents within the Earth produce its magnetic field through a process referred to as the geodynamo.

To determine the Earth's interior structure, seismic waves from earthquakes act as probes as they pass through the Earth. They reveal that the Earth's interior consists of three major zones or layers: the crust, mantle, and core. The crust and underlying mantle are composed mainly of rocky material, which is a good electrical insulator. The innermost region, the core, is composed of metals, most probably iron with a small percentage of nickel and an even smaller percentage of other elements, and thus it is a good electrical conductor. The inner core (out to a radius of about 1,300 kilometers) is solid, but the outer core (from 1,300 to about 3,500 kilometers radius) is molten.

The temperature of the solid inner core is estimated to be about 5,800 kelvins. Heat flowing outward through the outer molten core sets up convection currents, in which hotter, less dense fluid rises. When it transfers its heat to the overlying mantle, the fluid cools, becomes denser, and sinks. The convection may also be partly driven chemically. If iron crystallizes at the bottom of the molten outer core to add to the solid inner core, the remaining fluid contains less iron and so is less dense, augmenting the thermally driven upward motion. Simple convection currents are deflected by the Earth's rotation in a process called the Coriolis effect. Computer models of the outer molten core that incorporate both convection and rotation show that the fluid moves in a number of spiraling columns aligned roughly parallel with the Earth's rotational axis.

All that is needed to "jump start" the geodynamo is a weak background magnetic field, perhaps provided by the solar wind or a remanent primordial field. As the metallic fluid moves through the background field, electrical currents are induced that in turn generate their own magnetic fields. This produces a positive feedback that reinforces the electrical currents and the overall magnetic field. The approximate alignment of the fluid's spiral motion with the rotational axis produces a dipole field with the magnetic poles near the rotational poles. The energy to produce the stronger magnetic field comes from the motion of convection and rotation. This process does not continue generating an ever-increasing field; it levels off since it becomes harder to generate an even stronger field as the field strength increases.

The geodynamo process explains many features of the Earth's magnetic field. Although the magnetic poles apparently remain close to the rotational poles, there is a shift in the position of the magnetic poles and changes in the field strength over periods of years to centuries; this could be caused by changes in the convection currents within the molten outer core.

Furthermore, paleomagnetic studies indicate that the Earth's magnetic field has reversed polarity at irregular intervals many times in the geologic past, the last reversal occurring about 700,000 years ago. The geodynamo process is unstable over long periods of time and can decay and regrow with changed polarity. Geologists have constructed models of dynamos that are simple versions of the geodynamo, and when set in operation, these models display changes in the field's intensity and polarity. The geodynamo explains the origin of about 90 to 95 percent of the Earth's magnetic field; the rest probably comes from fields associated with magnetic minerals in the Earth's crust, more complicated irregularities in the convective motions of the molten outer core, and external sources such as the solar wind interacting with the Earth's ionosphere.

Methods of Study

The Earth's magnetic field is generated in its interior. A clue to the composition of the interior is provided by the Earth's average density. Dividing the Earth's mass by its volume shows the average density to be about 5.5 times the density of water. Common rocks from the surface are about 3 times denser than water. Therefore a portion of the Earth's interior must be much denser than surface rocks in order to yield the average value. Only metals have the required density, but some metals, such as aluminum, are too low in density, and others, such as uranium, are too high. Still others, such as gold and silver, are close to the required density but are too rare. Iron is a good candidate, since it has the right density and is fairly abundant.

The Earth's interior structure can be probed using seismic waves produced by earthquakes. Body waves from the earthquake travel through the Earth's interior. Their speed and direction of travel are determined by the density and elastic properties of the material through which they are traveling. The waves also are reflected off boundaries between different layers. When the transmitted and reflected waves reach the surface, they are recorded on seismographs. Analysis of the seismograms obtained at seismic stations all around the globe reveals that the Earth has three main zones or layers: the surface crust,

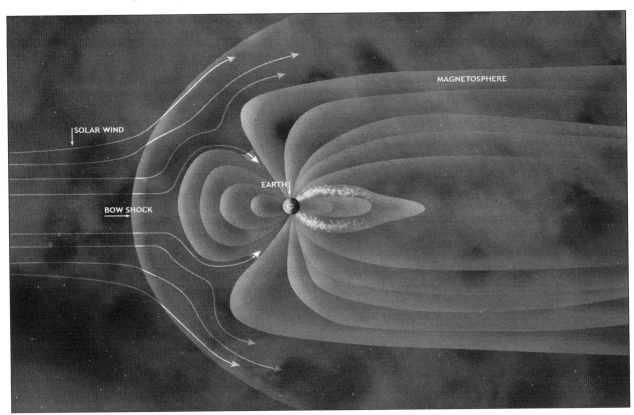

When the solar wind meets the Earth's magnetic field, a "bow shock" effect results, and the hot gases and radiation of the solar wind are deflected. Electrons are channeled by the magnetic field to create auroras during periods of high solar activity. (NASA/CXC/M. Weiss)

the mantle, and the central core. One type of body waves, called S waves, are transverse waves that can travel only through solids, not liquids. Their presence or absence recorded on seismograms reveals which parts of the interior are solid or liquid. The crust and mantle are solid (except for isolated pockets of molten material called magma). The inner part of the core also is solid, but the outer core is molten.

Models of the Earth's interior that combine chemical composition estimates with seismic-wave data indicate that the crust and mantle are composed of rocky material, while the core (both molten outer and solid inner parts) is mostly iron, with some nickel and other elements. The composition and pressure from these models can be used to calculate the melting-point temperature as a function of depth. The mechanical properties of the layers (whether solid or liquid, as indicated by seismic waves) can then be used to determine whether the actual temperature is below or above the calculated melting-point temperature. Anchored by measurements of heat flow at the surface and the increase of temperature with depth recorded in mines and wells, the geotherm—a graph of actual temperature versus depth—can be drawn. This is how the temperature of the Earth's core, about 5,800 kelvins, is determined, showing that there is enough heat energy to drive the convection necessary for the geodynamo process.

Remanent magnetism, the evidence of past magnetic fields preserved in some igneous and sedimentary rocks, can be measured with various types of magnetometers. These data, together with records of changes in the Earth's magnetic field during historic times, show how the field varies in strength and orientation and has even reversed polarity many times in the past.

Context

The geodynamo mechanism that generates the Earth's magnetic field also can explain the presence or absence of magnetic fields for other planets, their moons, and the Sun. Mars and Earth's moon have extremely weak magnetic fields, probably because their iron cores are so small, and they may have cooled to the point that their iron cores are no longer molten so convection cannot occur. Venus also has an extremely weak field; although it is nearly the same size and mass as Earth and probably has a similar internal structure with a molten outer core, it rotates very slowly.

Mercury has a magnetic field about an order of magnitude stronger than the fields of Venus, Mars, and the Moon, but about two orders of magnitude weaker than the

field of the Earth; although Mercury rotates slowly, its relatively large iron core might still have a molten convective zone. Jupiter's magnetic field is more than ten times stronger than Earth's, and Saturn's is about two-thirds as strong as Earth's; their fields are probably due to convection occurring in the liquid metallic hydrogen in their interiors (Jupiter has much more than Saturn) and their rapid rotation. Jupiter's moons Europa and Ganymede also have small magnetic fields, probably due to convection in electrically conductive salty oceans beneath their icy crusts. The Sun has a strong magnetic field that reverses polarity every eleven years; it is produced by the movement of ionized gas in the convection zone of the Sun's interior.

Stephen J. Shulik

Further Reading

Busse, F. H. "Recent Developments in the Dynamo Theory of Planetary Magnetism." In *Annual Review of Earth and Planetary Sciences* 11 (May, 1983): 241-268. An outline of the dynamo theory is given, with models of the dynamo for various planets. Observational evidence is discussed, along with the paleomagnetic data, geomagnetic reversals, and secular variation. Includes references and figures.

Fowler, C. M. R. *The Solid Earth: An Introduction to Global Geophysics*. 2d ed. New York: Cambridge University Press, 2004. An updated version of a widely used textbook for introductory geophysics courses. Designed for students with some knowledge of physics and calculus.

Garland, G. D. *Introduction to Geophysics*. 2d ed. Philadelphia: W. B. Saunders, 1979. Used as a text for introductory geophysics, this book covers, in sections 17.4 and 17.5, the cause of the main field and the dynamo theory. A few equations, but many figures and graphs that are of interest to the less technically informed reader. At the end of the chapter is a listing of thirty-two references.

Gubbins, D., and T. G. Masters. "Driving Mechanisms for the Earth's Dynamo." In *Advances in Geophysics*. Vol. 21, edited by B. Saltzman. New York: Academic Press, 1979. This article looks at such topics as the physical and chemical properties of the core and energy sources for the magnetic field. References are located at the end of the article. Mathematics and numerous figures and tables are included.

Hartmann, William K. *Moons and Planets*. 5th ed. Belmont, Calif.: Thomson Brooks/Cole, 2005. A college textbook beyond the introductory level, its approach is based on comparative planetology. Has a section on the generation of planetary magnetic fields by the dynamo mechanism.

Lapedes, D. N., ed. *McGraw-Hill Encyclopedia of Geological Sciences*. New York: McGraw-Hill, 1978. Pages 704-708, under the heading "Rock Magnetism," provide a concise description of many aspects associated with rock magnetism: how rock magnetization occurs, the present field, magnetic reversals, field generation, secular variation, and apparent polar wandering, among other subjects. Very readable, with no mathematics and a fair number of graphs, tables, and figures.

Merrill, R. T., and M. W. McElhinney. *The Earth's Magnetic Field*. New York: Academic Press, 1983. The authors cover much of the material associated with the Earth's field. Chapters 7 and 8 deal with the origin of the field, and chapter 9 covers the origin of secular variation and field reversals. Mathematical equations and thirty-eight pages of references. Numerous tables and figures.

Motz, L., ed. *Rediscovery of the Earth*. New York: Van Nostrand Reinhold, 1979. As a collection of articles for the nonscientist by scientists renowned in their respective fields, the text makes very interesting reading, augmented with many colorful illustrations. The chapter "The Earth's Magnetic Field and Its Variations" is written by Dr. Takesi Nagata, who has authored hundreds of articles on diverse aspects of geophysics besides the Earth's magnetic field, and covers a wide range of magnetic field topics. Two pages are devoted to the origin of the field. Includes a small amount of mathematics and only a few references.

Smith, D. G., ed. *The Cambridge Encyclopedia of Earth Sciences*. New York: Crown, 1981. Chapter 7, "The Earth as a Magnet," contains a discussion of the field's origin. The text is well written at a nontechnical level, with many colorful diagrams and figures.

Stacey, F. D. *Physics of the Earth*. New York: John Wiley & Sons, 1977. Under section 8.4, "Generation of the Main Field," the author provides a short, technical description of the origin of the field, which will be of interest to the more advanced student. Equations are rather formidable, but several figures illustrating the dynamo effect are included. A large number of references at the end of the text. Many other areas of geophysics are covered at a technical level.

Tarbuck, Edward J., and Frederick K. Lutgens. *Earth: An Introduction to Physical Geology*. Illustrated by Dennis Tasa. 9th ed. Upper Saddle River, N.J.: Pearson Prentice Hall, 2008. This college-level textbook for introductory geology courses is well written and illustrated. The chapter on the Earth's interior has a section on the generation of the magnetic field by the geodynamo process. The chapter on plate tectonics has a section on geomagnetic reversals.

Vogel, Shawna. *Naked Earth: The New Geophysics*. New York: Plume, 1996. Covers geophysics research and theories since 1960. Includes information about Pangaea, the supercontinent cycle, and the reversals of the Earth's magnetic field. For general audiences.

EARTH'S MAGNETIC FIELD: SECULAR VARIATION

Category: Earth

At every point on the Earth, its magnetic field has a direction, as indicated by a compass needle free to pivot three-dimensionally, and an intensity or strength. The direction and intensity of the magnetic field change over timescales of years to millennia, a phenomenon known as secular variation. Over longer geologic timescales of tens of thousands to millions of years, the Earth's field reverses polarity, a phenomenon called geomagnetic reversal.

Overview

Secular variation of the Earth's magnetic field refers to changes in the field's direction and intensity, a phenomenon manifested everywhere on the Earth's surface. Its most obvious effect is a gradual shift in the direction which an ordinary compass needle points. It also is seen in changes of inclination, the angle at which a magnetic needle suspended by its center tilts below the horizontal, as well as variations in the field intensity or strength. These changes appear to be noncyclic and occur over timescales of years to millennia.

The direction of the Earth's magnetic field at any point is specified by two angles called declination and inclination. The north end of a magnetic compass needle points approximately to the north, but not exactly. The angle between geographic or true north (defined by the Earth's north rotational pole) and magnetic north originally was called magnetic variation, but now is called declination. It was first noticed around the twelfth century, when it was thought to be caused by abnormalities in the compass needle's magnetization or suspension. However, by the early sixteenth century, Europeans had accepted declination as a phenomenon of the Earth's magnetism. Inclination, the downward tilt of a compass needle free to pivot three-dimensionally, was also discovered during that century. Hence William Gilbert could write in 1600 of both declination and inclination as natural features of the Earth's magnetism.

By the early sixteenth century, Europeans had noticed that the declination varies from place to place (and this helped convince them that declination was a feature of the Earth's field and not due to flaws in their compasses). The discovery arose in the practices of navigation, chart making, and crafting of magnetic compasses—all activities connected with exploration. Perhaps Christopher Columbus, and certainly Sebastian Cabot, noted that while compass needles pointed east of north near Europe, they pointed west of north in the New World.

All three measures of the magnetic field—declination, inclination, and intensity—vary over the entire planet. These variations are most easily depicted with maps on which curved lines connect points that have the same value of one of these three parameters. For example, maps that display curved lines of equal magnetic declination are called isogonic maps. The first such printed map was produced in about 1701 by Edmond Halley of comet fame. Initially it was hoped that isogonic maps could be used to determine longitude by compass. Maps similar to isogonic maps but displaying lines of equal inclination or equal intensity also can be drawn.

Meanwhile, Henry Gellibrand announced in 1635 that declination changes over time as well as space. He found that the magnetic declination for London had shifted from 11.3° east of north in 1580 to 4.1° east of north in 1634. Later investigators discovered that the inclination and the intensity of the magnetic field also gradually change. For example, between 1700 and 1900, the inclination at London decreased from about 75° to 67°. Currently the overall intensity of the dipole field is decreasing at the rate of about 6 percent per century. If the field were to continue to decrease at this rate, it would drop to zero in about sixteen hundred years.

The agonic line is the "line of zero declination," along which compass needles point exactly toward geographic or true north. One aspect of secular variation has been the westward drift of the agonic line. This drift can be depicted on maps; just as one can map the magnetic parameters, one can also map how these parameters change, by drawing curved lines connecting points that change at the same rate. For example, all points where the declination is shifting westward at 10° per century would be connected together. These charts, known as isoporic charts, came into wide use in the mid-twentieth century. Areas of most rapid change are called isoporic foci. These isoporic foci are drifting westward, just as the agonic line is. While this westward drift has been a persistent feature of secular variation since Gellibrand, first pointed it out, some evidence exists for eastward drifts during prehistoric times.

Geologic evidence of ancient magnetic fields preserved in some igneous and sedimentary rocks shows that the geomagnetic field has reversed polarity many times over intervals of tens of thousands to millions of years. These geomagnetic reversals have played a major role in plate tectonics. They provide evidence of seafloor spreading and can be used to determine its rate. They indicate the locations of continents in the geologic past and thus can be used to trace continental drift.

The geomagnetic field and its secular changes traditionally have been attributed to the Earth's interior. Thus theories of the source of the field and the causes of its secular variation are necessarily indirect and extremely diverse, given the inaccessibility of the interior for direct study. About four hundred years ago, Gilbert suggested that the Earth behaves as if it had a bar magnet or magnetic dipole of extraordinary intensity at its center. In 1674, Robert Hooke asserted that the magnetic dipole axis of the Earth is tilted about 10° from the axis of rotation and that the dipole axis rotates westward around the rotational axis every 370 years. In 1683, Halley proposed that a double dipole pattern with four magnetic poles provided a better fit to worldwide declination data than just two magnetic poles, and in 1692, he suggested that his four poles could explain secular variation. Two of these poles he assigned to the Earth's outer crust and the other two to a central nucleus, which rotated slightly more slowly than the crust, on the same axis. The crustal magnetic poles were fixed in place, and as the nucleus, rotating a bit more slowly, drifted slowly westward relative to the crust, so did its magnetic poles. This explained, he thought, the drift of the agonic line.

Theories that the core is permanently magnetized were later ruled out when models of the Earth's interior showed that it is too hot; its temperature is above the Curie temperature of all known permanently magnetized materials. The Curie temperature (or Curie point) is the temperature above which a material is no longer permanently magnetic.

It is known that moving electrical charges generate magnetic fields. In particular, an electric current flowing around a wire loop produces a magnetic dipole field through the center of the loop. During the nineteenth and twentieth centuries, theories were developed that attributed the magnetic field and its secular variations to electric currents in the Earth's interior. One hypothesis was that the rotation of the Earth's iron core carried the charges with it, and this motion generated the field. This theory reached its highest state of development around 1950 in work by Patrick M. S. Blackett, but since then it has gradually lost favor. An alternative hypothesis is that the flow of molten metal in the interior carries the charge that generates the field. First introduced in rudimentary form in the nineteenth century, this theory became increasingly sophisticated with the investigations of Walter Elsasser beginning in 1939 and Sir Edward Crisp Bullard starting in 1948. Elsasser proposed that the combination of the movement of molten metal and the simultaneous flow of electricity in it produced both the Earth's main dipole field and its secular variations. This dynamo was driven, he suggested, by the heat generated by the decay of radioactive materials in the interior. Convection of hotter, less dense materials upward and of colder, denser materials downward, he said, produced the dynamo.

There now is general agreement that the geomagnetic field and its secular variation are the result of fluid motions in the molten outer part of the Earth's metallic core. Various models have been developed to show that convective motion in the molten outer core, modified by the Coriolis effect due to the Earth's rotation, can produce the observed field and its secular variations. Metals are good electrical conductors because electrons can move easily through metals. The molten metal of the outer core, as it moves through the magnetic field, makes electrons in the metal move, inducing electric currents in the metal that in turn generate the magnetic field. Thus the geodynamo is self-sustaining, but it is not a perpetual motion machine; it needs an energy source to drive the motion. The geodynamo does not create its magnetic field from nothing; rather, it converts some other form of energy into magnetic energy. The two most probable energy sources to drive the convection and generate the field are heat from the decay of radioactive materials and the crystallization and settling of iron (and other dense metals) to the solid inner core.

In the end, one must remember that models of the geodynamo and its energy sources are tentative. Many

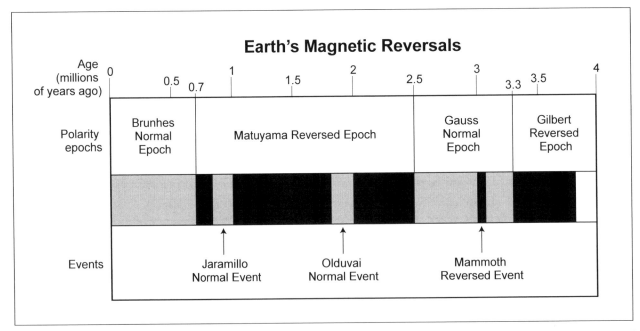

Geologic evidence of ancient magnetic fields preserved in some igneous and sedimentary rocks shows that Earth's geomagnetic field has reversed polarity many times over intervals of tens of thousands to millions of years.

debates are still waged over the details of the geodynamo and how it produces secular variation. This area of geophysical theory is a most active and challenging one, and it is in rapid flux.

Methods of Study

The simplest way to detect secular variation is to observe the changing declination of a magnetic compass over some decades; until the twentieth century, that was the only way. All the instruments employed by famous investigators of geomagnetism, from Gilbert in 1600 to Carl Friedrich Gauss in the 1830's, used adaptations of the compass to measure the magnetic parameters and their changes. Among other goals, these scientists aimed to measure these elements more accurately, so as to reveal secular change in shorter time intervals. During the twentieth century, however, there was a sustained trend to replace traditional magnetic needle instruments with ones based on other applications of electromagnetic principles.

Around 1900, research-quality Earth inductors were developed to replace the dip needle and circle in measuring inclination. The idea behind this first of the new electrically based geomagnetic instruments is simple. Rotate a coil of wire about its own diameter in a magnetic field. If the rotational axis differs from the direction of that field, an electric current is induced in the coil, but if the axis coincides with the field, the current will cease. This "null" method now is used to measure inclination more accurately and easily.

The Earth inductor was followed in the 1930's by the flux-gate magnetometer. This instrument is based on a high-permeability alloy, that is, one that magnetizes readily. Around two cores of such material are wound two coils of wire, in opposite directions, that carry the same alternating current, so that the same reversing magnetic field is produced in both cores, but 180° out of phase. When placed in the Earth's field, the changes in the magnetic fields of these two cores do not cancel out, and this changes the current flowing in the coil around each core by different amounts. The net current is related to the component of the Earth's magnetic element in the direction the magnetometer is pointing. When, however, this magnetometer is oriented parallel to the Earth's field, no current is produced. The flux-gate magnetometer has seen wide use in aerial geomagnetic surveys.

Other generations of magnetic instruments have appeared since the flux-gate. Some of the most useful are proton precession magnetometers, rubidium vapor magnetometers, and superconducting magnetometers. These devices take advantage of principles of quantum physics. Some of them, like the proton precession instrument, measure only the total intensity of the Earth's field. Others, like the superconducting magnetometer, are directional. Both types are many times more sensitive than older instruments and also perform much faster.

Magnetic surveys have been an essential part of the method of studying secular variation. All over the world, teams of observers have established "repeat stations," or places for careful observation of the magnetic field parameters at various time intervals. Magnetic surveys have been greatly facilitated not only by the new instruments mentioned above but also by the way those instruments are used. Surveys are now often conducted very quickly with instruments carried by airplanes and satellites (such as MAGSAT). Data that once took decades to gather are now collected in months. Moreover, the extensive calculations needed to analyze global data have been greatly accelerated by computers. Worldwide magnetic charts are produced much more frequently now than in 1900, and the study of secular variation is thus much more detailed.

Equally impressive changes have been wrought by the use of geomagnetic methods to study the magnetic properties of rocks. Igneous and sedimentary rocks that contain iron grains can record the Earth's magnetic field at the time they formed. The phenomenon is called remanent magnetism or paleomagnetism. Until the development in the middle of the twentieth century of techniques for measuring remanent magnetism in rocks, secular variation studies were limited to data obtained by direct measurement of the Earth's field during historic times. Little was known of the magnetic field before 1600. Past phenomena that have been revealed by these methods include reversals of the magnetic field polarity and geomagnetic excursions.

The study of geomagnetism has come a long way with the rapid development of new instruments and methods. No longer is the purpose restricted to just a description of the main field and its variations. With the new sensitivity and portability made possible by electronics, geomagnetic secular variation has become a useful tool in many diverse scientific endeavors, such as archaeological dating of artifacts, magnetostratigraphic dating of sediments, determining rates of seafloor spreading, and tracing continental drift, in addition to the traditional effort to understand processes occurring in the Earth's core.

Context

Most people are familiar with the magnetic compass, and many know at least roughly how to use it. Two activities which demand close attention to magnetic declination and its secular variation are reading topographic maps and navigating at sea.

In the margin of topographic maps, there usually are arrows which point to true north and magnetic north. With this declination information, one can relate directions on the map to compass readings in the field. In areas where the secular variation of declination occurs rapidly, it is also necessary to know when declination readings were last measured and the rate of their change. For example, near Tay River in the Canadian Yukon, the declination was listed as 33°, 25 arc minutes, east of north in 1979 and decreasing at 3.3 arc minutes per year. Thus, if the secular variation there continued at that rate, in a century the declination would change by 5°, 30 arc minutes, to 27°, 55 arc minutes east of north. Secular variations cannot be predicted reliably over so long a period, however, and maps are therefore updated regularly in magnetic surveys.

Information regarding declination at sea and especially near the coast is of even greater importance. Every ship is sometimes beset by fog, and thus an essential bit of navigational data is the present declination. Up-to-date charts are, again, the best means to avoid dealing with secular variation. As the date of the magnetic declination recedes into the past, however, reliable information concerning its secular change becomes more important.

The deep interior of the Earth is inaccessible to direct study. Thus scientists must watch closely for clues received at the Earth's surface about the conditions and processes in the interior. Magnetic secular variation is one of the ways information can be obtained about the geodynamo. The geomagnetic changes occurring over geologic time have been preserved in some igneous rocks as they cooled and some sedimentary as they settled polarity reversals of the Earth's main magnetic field. This remanent magnetism or paleomagnetism records what the field was like in the past. Such data gathered around the world from rocks of various ages reveal that the geomagnetic field has reversed its polarity many times in the geologic past. These geomagnetic reversals typically occur at intervals of tens of thousands to millions of years; the last one happened about 700,000 years ago. The geomagnetic reversals can be tied in to the chronology of the Earth and are an important element in plate tectonics.

Gregory A. Good

Further Reading

Backus, George. *Foundations of Geomagnetism*. New York: Cambridge University Press, 1996. Describes in detail the mathematical and physical foundations of geomagnetism. Technical and more advanced than introductory texts.

De Bremaecker, Jean-Claude. *Geophysics: The Earth's Interior*. New York: John Wiley & Sons, 1985. This well-written text is intended for college-level students with some calculus and some physics background. Nevertheless, the author is careful to explain difficult concepts or mathematical statements. Chapter 10, "Magnetostatics," and chapter 11, "The Earth's Magnetic Field," can be read separately from the rest of the book to provide an in-depth survey of geomagnetism, its measurement, and its secular variation. Especially useful are the technical appendixes on mechanical quantities, magnetic quantities, data about the Earth, notation, and some relevant mathematics. One of the best treatments available.

McConnell, Anita. *Geomagnetic Instruments Before 1900*. London: Harriet Wynter, 1980. This short book provides one of the clearest expositions of the basics of geomagnetism for the lay reader. Includes illustrations of many of the basic early forms of instrumentation, especially European.

Tarbuck, Edward J., and Frederick K. Lutgens. *Earth: An Introduction to Physical Geology*. Illustrated by Dennis Tasa. 9th ed. Upper Saddle River, N.J.: Pearson Prentice Hall, 2008. This college-level textbook for introductory geology courses is well written and illustrated. The chapters on the Earth's interior and plate tectonics have sections on the Earth's magnetic field and its reversals in the geologic past.

Thompson, Roy, and Frank Oldfield. *Environmental Magnetism*. London: Allen & Unwin, 1986. This book captures the broad range of possible applications of knowledge of magnetism in the study of the Earth that have appeared since the 1950's, from the study of magnetic minerals to biomagnetism. This is an introductory, nonmathematical, college-level text. Although its chapters on basic magnetic principles are valuable, the most unusual feature of the book is the many application chapters. Especially relevant to secular variation are chapters: "The Earth's Magnetic Field"; "Techniques of Magnetic Measurements"; "Reversal Magnetostratigraphy"; and "Secular Variation Magnetostratigraphy."

EARTH'S MAGNETOSPHERE

Category: Earth

The Earth's magnetosphere is the region around the Earth in which the geomagnetic field is stronger than the interplanetary magnetic field. The outer boundary of the magnetosphere is termed the magnetopause. The flow of charged particles from the Sun, called the solar wind, pushes in the magnetopause on the side of the Earth facing the Sun in a bow shock effect and drags it out into a long magnetotail on the side facing away from the Sun. The Earth's magnetic field interacts with the charged particles of cosmic rays as well as those emanating from the Sun, protecting life from their harmful effects.

Overview

The magnetic field at Earth's surface has been used in navigation for centuries. However, studying the extent and behavior of the geomagnetic field out in space around Earth became possible only with the dawn of the space age. The launch on January 31, 1958, of Explorer 1, the first successful U.S. satellite, brought a completely unexpected discovery. Belts of electrically charged particles circling Earth and trapped by Earth's magnetic field were detected by James Van Allen when he analyzed the signals sent back from a Geiger counter he had placed on board the satellite. These belts were named the Van Allen radiation belts in his honor. Since Explorer 1, myriad spacecraft have explored Earth's magnetosphere, measuring the field's strength and direction, the numbers and energies of the charged particles in it, and its interaction with the solar wind and interplanetary magnetic field.

The magnetic field near Earth is basically a dipole field, the type of field produced by a bar magnet or an electric current flowing in a wire loop. It is as if a bar magnet were inside Earth, tilted about 10° to 15° with respect to the axis of rotation and offset from Earth's center several hundred kilometers toward the Pacific Ocean. The north pole of this hypothetical magnet actually is located in the Southern Hemisphere and is called the south magnetic pole; the south pole of the magnet is located in the Northern Hemisphere and is called the north magnetic pole. The north magnetic pole (actually a magnetic south pole) is located north of Canada and west of Greenland about 8° away from the geographic or rotational north pole, and the south magnetic pole (actually a magnetic north pole) is located off the coast of Antarctica in the direction of Australia about 26° from the geographic or rotational south pole. This somewhat confusing situation and terminology is because opposite poles attract, and by definition a magnet's north and south poles are the north-seeking and south-seeking ends of the magnet used as a compass.

Although imagining a bar magnet inside Earth is an easy way to visualize the geomagnetic field, this cannot be the actual situation, since Earth's interior is much too hot for any material to be permanently magnetic. Instead, the source of the geomagnetic field and magnetosphere is thought to be fluid motions and electric currents in Earth's

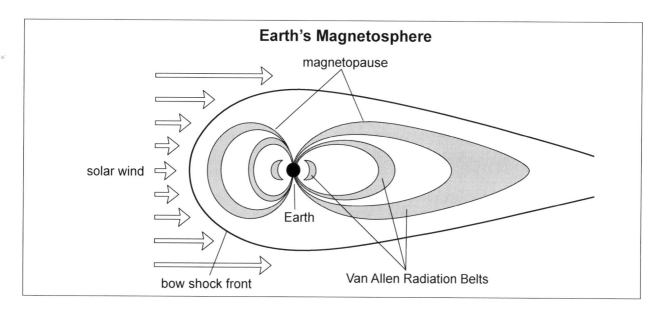

Earth's Magnetosphere

magnetopause

solar wind

Earth

bow shock front

Van Allen Radiation Belts

molten metallic outer core, in what is termed the geodynamo process.

With increasing distance from Earth, the geomagnetic field lines become distorted from those of a simple bar magnet by the solar wind—the flow of electrically charged particles blown from the Sun's surface out into interplanetary space. The flow of charged particles carries the solar magnetic field, which becomes the interplanetary magnetic field. The solar wind pushes back the geomagnetic field lines on the side of Earth facing the Sun and stretches them out into a magnetotail of nearly parallel field lines on the side away from the Sun. The boundary where the geomagnetic field equals the solar/interplanetary magnetic field is called the magnetopause; inside it is Earth's magnetosphere, where the geomagnetic field dominates. The magnetosphere typically extends out to about 65,000 kilometers on the side of Earth facing the Sun and up to millions of kilometers in the magnetotail pointing away from the Sun.

Where the rapidly moving charged particles of the solar wind encounter the magnetosphere, their speed abruptly decreases and their density abruptly increases, forming a shock wave called a bow shock, analogous to the water-wave pattern around the bow of a rapidly moving boat. Most of the particles in the solar wind flow past Earth around this bow shock and drag the magnetosphere out into its long magnetotail. The magnetopause and its accompanying bow shock fit around Earth like a sock with a golf ball in its toe.

Some of Earth's field lines originating near a magnetic pole extend through the magnetopause into the magnetosheath, the region between the magnetopause and the bow shock. These field lines form the polar cusps or clefts where they pass through the magnetopause. It is along these field lines that some of the charged particles in the solar wind can enter Earth's magnetosphere through the polar cusps.

The magnetic field lines that exit the surface in one hemisphere and bend around Earth and reenter the surface in the other hemisphere are called closed field lines. Those that do not return but extend through a polar cusp into the magnetosheath or continue out into the magnetotail are called open field lines. At the magnetic poles, the boundary between the open field lines and the closed field lines is called the auroral oval. Because the auroral oval is the dividing line between closed field lines where particles may be trapped and open field lines where trapping is impossible, it is where the energetic charged particles of the solar wind can enter Earth's atmosphere and produce auroras. Spacecraft images have shown that the auroras are fairly continuous phenomena around the auroral ovals. Charged particles also can escape through the auroral ovals; in particular, streams consisting mainly of helium ions—the polar wind—flow out to distant regions of the magnetosphere.

Along the central plane of the magnetotail runs a flattened plasma sheet of ions (atoms that have lost one or more electrons) and electrons, separating the magnetic field lines originating from Earth's north and south magnetic poles. Currents flow crosswise in the plasma sheet and maintain the magnetic field separation. At great distances from Earth, the magnetotail may sometimes undergo a magnetic field line reconnection or collapse. This releases a large amount of stored magnetic field energy, which in turn accelerates the charged particles in the magnetosphere. Such an event is called a geomagnetic substorm. During a substorm, high-energy electrons enter the atmosphere near the auroral oval, producing particularly strong auroras and disrupting the ionosphere and long-range radio communications, which are reflected by the ionosphere. At these times, ring currents flow around the magnetic equator at a distance from Earth of about 12,500 kilometers. Magnetometers on the ground can measure changes in the magnetosphere produced by these currents.

At times of increased solar activity, near the maximum of the solar sunspot cycle, solar flares send out strong bursts of charged particles in the solar wind, which can strongly compress the magnetosphere on the Sun side of Earth and collapse the magnetotail on the far side. Then the magnetopause may shrink to within 12,000 kilometers of Earth. When that occurs, geosynchronous satellites orbiting at a distance of about 40,000 kilometers (where the satellite revolution period equals Earth's rotation period) enter the magnetosheath and become subject to interactions with solar wind particles, causing the satellites to charge suddenly to high voltages. Some communications satellites have been disabled or have had their signals disrupted by internal sparks at such times.

When charged particles such as electrons or ions move in a magnetic field, they follow corkscrew paths around the magnetic field lines. As a charged particle corkscrews around a magnetic field line, the angle its path makes with the field line (its "pitch angle") increases as it moves into regions of greater magnetic field strengths, until it is moving in a circle perpendicular to the field line and can go no farther in the direction of the field. It has reached its "mirror point" and must now return in the direction from which it came.

The field lines of a magnetic dipole reach out from regions of high field strength near one pole to regions of weaker intensity at the magnetic equator, and then return to regions of high field strength near the opposite pole. Consequently, a particle corkscrewing around one of these field lines will be mirrored near one magnetic pole, travel back to the vicinity of the other pole and be mirrored again, and so on. High-energy particles can reach farther in toward the pole than low-energy particles, and they may even enter the atmosphere, producing the beautiful auroras, or northern and southern lights. Less energetic particles are trapped on the magnetic field lines outside the atmosphere, and they can only escape by colliding with other particles or by slowly moving across field lines in response to electric fields or changing magnetic fields. As Earth rotates, these trapped particles swing around with the rotating magnetic field.

Knowledge Gained

Earth's magnetosphere is made up of Earth's magnetic field and the charged particles controlled by it. The magnetosphere and magnetic field can change over hours, days, or years in response to currents of charged particles and the surrounding interplanetary magnetic field.

Earth has two invisible belts of trapped energetic particles, the Van Allen belts. They were discovered by James Van Allen in 1958 using data from the first successful U.S. satellite, Explorer 1. The inner belt, which extends from 1,000 to 5,000 kilometers above the equator, is kidney-shaped in cross section and contains mainly high-energy protons. The outer belt, 15,000 to 25,000 kilometers from Earth, is crescent-shaped in cross section and contains mainly high-energy electrons. Because the charged particles are trapped in the magnetic field, they cannot easily leave the belts. They pose a danger to astronauts or sensitive electronic equipment orbiting for long periods within the Van Allen belts.

Context

It is important to understand the processes that occur in Earth's magnetosphere because they strongly affect conditions on Earth. Only since the space age began have the structure and processes of Earth's magnetosphere been subject to study.

Earth's magnetosphere performs the vital function of shielding living things from the possibly harmful effects of high-energy charged particles, such as cosmic rays and those that come from the Sun. When space travelers leave the protection of the magnetosphere, they must be shielded against charged particles, especially the intense bursts of them emitted in solar flares. Furthermore, strong solar flares may disrupt the magnetosphere, damaging or disabling Earth-orbiting satellites, long-distance radio communications, and electricity distribution networks.

At times in Earth's history when its magnetic field became very weak and reversed direction, living things may have been subjected to an intense flux of high energy charged particles, from cosmic rays and solar flare ejections, which could have led to increased rates of cancer and genetic mutation. Genetic mutations can be both good and bad. Although most mutations are deleterious (and even deadly), some are beneficial and convey an evolutionary advantage. Therefore, those times when the geomagnetic field was weak or absent may have contributed to mass extinctions or rapid evolution.

Some theorists propose that changes in the number of charged particles entering Earth's atmosphere, or changes in the solar/interplanetary magnetic field as it interacts with Earth's magnetic field, may influence the weather on Earth. Perhaps, they suggest, auroral currents modify polar air currents, leading to periods of drought or increased precipitation in different regions on Earth.

Dale C. Ferguson

Further Reading

Akasofu, S. I., ed. *Dynamics of the Magnetosphere.* Dordrecht, Netherlands: D. Reidel, 1980. A collection of contributions to a 1979 meeting of magnetospheric scientists in Los Alamos, New Mexico. Includes results of experiments and readable, condensed, technical summaries. This volume is mostly concerned with the disturbed magnetosphere, as during geomagnetic substorms. Well illustrated.

Akasofu, S. I., and Y. Kamide, eds. *The Solar Wind and the Earth.* Dordrecht, Netherlands: D. Reidel, 1987. Written by experts, this is a collection of chapters about the Sun and Earth and their interactions. Includes sections on Earth's ionosphere and thermosphere. The book is somewhat technical, but nevertheless clear; the history of each subtopic is well treated. Contains lists for further reading.

Allen, Oliver E. *Atmosphere.* Alexandria, Va.: Time-Life Books, 1983. A popular book that covers all aspects of the atmosphere. A good treatment of the history of atmospheric studies. The interested layperson will find the relationship of the magnetosphere to Earth's atmosphere well explained. Contains photographs, illustrations, and a bibliography.

Chaisson, Eric, and Steve McMillan. *Astronomy Today*. 6th ed. New York: Addison-Wesley, 2008. Very well-written college-level textbook for introductory astronomy courses. Has a section on Earth's magnetosphere.

Fraknoi, Andrew, David Morrison, and Sidney Wolff. *Voyages to the Stars and Galaxies*. Belmont, Calif.: Brooks/Cole-Thomson Learning, 2006. A well-written, thorough college textbook for introductory astronomy courses. Has a section on Earth's magnetosphere.

Friedman, Herbert. *Sun and Earth*. San Francisco: W. H. Freeman, 1986. A volume that lucidly describes the Sun's effects on Earth's magnetosphere and ionosphere. Written for the layperson by a pioneer in spacecraft exploration. Of moderate length, the book contains photographs, drawings, and an appendix of references to specific topics.

Hargreaves, John K. *The Upper Atmosphere and Solar-Terrestrial Relations*. New York: Van Nostrand Reinhold, 1979. This textbook delves into the physics of the magnetosphere. It may profitably be read by specialists in the field, other physical scientists, or mathematically inclined students. Includes useful line drawings, numerous equations and references, and questions for study.

Hartmann, William K. *Moons and Planets*. 5th ed. Belmont, Calif.: Thomson Brooks/Cole, 2005. A college textbook beyond the introductory level, its approach is based on comparative planetology. Provides a succinct but easily understood description of planetary magnetic fields and magnetospheres.

Johnson, Francis S., ed. *Satellite Environment Handbook*. 2d ed. Stanford, Calif.: Stanford University Press, 1965. An excellent technical reference that has lost little value with age. It covers the near-Earth environment, from the magnetic field to micrometeoroids, and has a good section on the magnetosphere. Includes graphs, tables, line drawings, and references. Written by many knowledgeable contributors.

Vogel, Shawna. *Naked Earth: The New Geophysics*. New York: Plume, 1996. Covers geophysics research and theories since 1960. Includes information about Pangaea, the supercontinent cycle, and the reversals of the Earth's magnetic field. For general audiences.

EARTH'S ORIGIN

Categories: Earth; The Solar System as a Whole

The Earth's early formation, its subsequent internal differentiation, its active plate tectonics, and its external weathering have left little substantive evidence of its origin intact for direct study. Much about the materials and formative processes involved in the planet's origin can be deduced, however, from seismology, geomagnetics, and the study of meteorites and comets.

Overview

In order to understand the origins of the Earth, it is necessary to be aware of the sources of the materials from which it is made. The matter from which the Earth and the entire universe is made was created in the big bang, about 13 to 14 billion years ago. The processes that occurred in the first few minutes after the big bang produced all the hydrogen and most of the helium in the universe today; trace amounts of lithium and beryllium also were formed.

Later, after stars formed, they produced all the other chemical elements through nuclear fusion reactions, also called nucleosynthesis. For most of their lives, stars generate the energy to shine by fusing lighter atomic nuclei together to make heavier atomic nuclei. This requires high temperatures and densities, the conditions that exist in the interiors of stars. The first step is the fusion of four hydrogen nuclei into one helium nucleus. The next step is the fusion of three helium nuclei into one carbon nucleus, maybe adding a fourth helium to produce oxygen. This is the end of energy generation and nucleosynthesis for a Sun-like star before it puffs off its outer layers as a planetary nebula, and the exposed core cools and fades to end its life as a white dwarf and ultimately as a black dwarf. (Note that planetary nebulae have nothing to do with planets or the formation of planets. The name dates back to the 1700's, when, viewed through telescopes of that time, they looked fuzzy, like nebulae, and round, like planets).

More massive stars have a more spectacular demise. After carbon and oxygen have formed, further fusion reactions continue to form heavier elements up to iron. The production of elements even heavier than iron does not generate energy but requires the input of energy. The iron core collapses, and the outer layers collapse on top of it and then rebound, tearing the star apart in a supernova explosion. The tremendous energy released in a supernova permits the formation of the rest of the chemical elements.

The chemical elements produced during the star's life and death are dispersed by the supernova explosion into interstellar space, there to enrich clouds of gas called nebulae (containing mostly hydrogen and some helium) in the heavier chemical elements.

The Sun, planets, and other bodies of the solar system formed as the result of the gravitational contraction of part of such a nebula about 4.5 to 4.6 billion years ago. The portion that would become the solar system, called the solar nebula, initially was perhaps about a light-year (about 9.5 trillion kilometers) across and was composed of about 74 percent hydrogen, about 24 percent helium, and about 2 percent all the other chemical elements. The exact mechanism responsible for the initiation of the solar nebula's contraction is still speculative. It may have involved the compression of the nebula as it passed through a spiral arm density wave as the nebula orbited the center of our galaxy, the Milky Way. It may have been triggered by a shock wave propagating through the nebula when a nearby massive star went supernova. It is generally agreed that once started, gravitational effects within the solar nebula kept the process going.

Any initial slow rotation of the solar nebula increased its speed with the contraction of the nebula to conserve angular momentum. (The same effect is seen on spinning figure skaters, whose rotational speed increases as they bring their arms close to their bodies.) The increase in rotational speed caused the nebula first to become oblate and eventually to form a flattened equatorial disk. Most of the solar nebula's mass concentrated at the center of the disk, forming the proto-Sun, which grew hotter by gravitational contraction.

As the proto-Sun formed at the center, fractional condensation began—a process in which gaseous matter solidifies into small, sand-sized grains only in regions where the ambient temperature is below the material's melting point. Only metallic grains of iron and nickel condensed close to the proto-Sun. Farther out, where the temperature was lower, they were joined by grains of silicate minerals. Still farther out, various ices of water, carbon dioxide, methane, and ammonia could condense. These solid grains collided with one another and stuck together in a process called accretion, forming planetesimals that grew in size. Within a time span of a few tens of millions to perhaps one hundred million years, the largest planetesimals grew into protoplanets, while the smaller ones became the many satellites and other minor members of the solar system.

As the protoplanets continued to grow, their gravitational influence grew as well. They could attract greater amounts of disk material, thus accelerating their growth while at the same time sweeping the surrounding interplanetary space clean. Solar radiation could then penetrate the space between the Sun and the planets, bringing light and heat to their still-evolving surfaces. It was during this time that the planets started to evolve in different ways. The third planet from the Sun, Earth, and its inner solar system neighbors (Mercury, Venus, and Mars) had relatively weak gravitational fields, which, coupled with their now high surface temperatures and exposure to the solar wind, caused them to lose significant amounts of the lighter gases. This first atmosphere, probably consisting of hydrogen, helium, methane, ammonia, carbon dioxide, and water vapor (gases common in the solar nebula), escaped from the inner planets and was blown away into the outer solar system. The outer planets, Jupiter and beyond—because of their colder temperatures, their greater masses, and consequently the lessened influence of the solar wind—were not so affected. As a result, they became the low-density gas/liquid/ice "giant" planets with small, rocky/metallic cores.

As the accretion process drew to a close, the Earth (along with the other inner planets) was subjected to a final intense bombardment of impacting planetesimals. As each colliding object struck the Earth's surface, its energy of motion was converted into heat energy. Furthermore, radioactivity levels were much higher in the very early Earth, since many of the radioactive elements with shorter half-lives had not yet decayed. As the Earth grew larger in size, it tended to insulate itself, making it more difficult for the energy released by radioactive decay in its interior to reach its surface and escape. All of these effects served to increase the early Earth's temperature to the point that it at least partially melted, allowing chemical differentiation and the development of the Earth's layered internal structure. Molten blobs of heavy metals like iron and nickel sank to the center, forming the iron-rich core. Less dense silicate and oxide minerals remained behind, forming the mantle. The surface also melted, forming a magma ocean perhaps a few hundred kilometers deep. Eventually the surface cooled and hardened into a thin, primitive basaltic crust, probably similar to present-day ocean-floor crust.

At the same time, outgassing released gases trapped in the interior at a prodigious pace, producing the Earth's second atmosphere, probably consisting mostly of water vapor, carbon dioxide, and sulfur dioxide, with smaller amounts of nitrogen, hydrogen sulfide, and other gases, but no free oxygen. (Outgassing continues today through volcanoes, fissures, and fumaroles, but at a much reduced rate.)

As the Earth cooled, water vapor in the atmosphere condensed, formed clouds, and fell as rain, forming the first streams and oceans. Due to the abundant carbon dioxide along with sulfur dioxide and hydrogen sulfide in the atmosphere, this early rain was highly acidic, resulting in rapid chemical weathering of surface rocks. The weathering products were carried by streams into the oceans, rapidly increasing their salinity. By about 4 billion years ago, the oceans had reached nearly their present volume and degree of saltiness. Large amounts of carbon dioxide from the atmosphere dissolved in the oceans, combined with other dissolved materials, and precipitated out as sediment (mostly as the mineral calcite, as calcium carbonate). As other gases were removed from the atmosphere, nitrogen remained, eventually becoming the major atmospheric constituent.

Probably by about 3.8 billion years ago, the first life appeared. Organic molecules, including amino acids, may have formed in the Earth's early atmosphere and oceans with solar ultraviolet light, lightning, or deep-sea hydrothermal vents providing the needed energy input, or they may have been delivered to the Earth's surface by impacts of comets, asteroids, and meteoroids. The earliest living organisms were anaerobic (able to survive without oxygen), since there was no free oxygen in the atmosphere and oceans. Then cyanobacteria and possibly other early organisms developed photosynthesis, using sunlight to turn water and carbon dioxide into sugars for food. This reaction released free oxygen into the oceans and atmosphere, and life evolved to utilize it to extract energy from food.

Thus the Earth was transformed from its formative stages to what it is today, a place where life thrives, where rocks are formed and weathered on the surface, and where a dynamic interior drives tectonic processes.

Methods of Study

Much of what is known about the origins of the Earth is derived by studying meteorites and comets as well as from seismology and geomagnetics. Meteorites and comets are unaltered or little-altered examples of early solar-system materials, providing information on the composition of and processes that occurred in the early solar nebula and the various types of bodies that formed from it. Seismology and geomagnetics give researchers clues about the internal structure of the planet.

Meteorites are extraterrestrial pieces of rock or metal that survived their fall through the Earth's atmosphere. The combined total composition of all meteorites is probably representative of the rocky and metallic material from which the inner planets—Mercury, Venus, Earth, and Mars—formed. Comets provide evidence of the more volatile components of the early solar system. They are composed of various "dirty" ices, indicative of materials blown out from the inner solar system by the solar wind but not before a portion was incorporated into the accreting planetesimals.

Data obtained from seismology (the study of the transmission of earthquake shock waves through the Earth) led to the discovery that the Earth's interior is divided into several distinct layers or zones. Observations of a change in speed of seismic waves near the Earth's surface led to the discovery of the Mohorovicic (Moho) discontinuity, the boundary between the crust and mantle. A low seismic velocity zone is now recognized in the upper mantle below the Moho and is used to define the lower boundary of the Earth's rigid lithosphere and the top of the "plastic," deformable asthenosphere. Another seismic discontinuity 2,900 kilometers below the surface delineates where the solid mantle is separated from the molten outer core. Later seismic work revealed the existence of a solid inner core.

Studies of the magnetic field of the Earth also give support to the zonal nature of the planet's interior. Hypotheses concerning the generation of the magnetic field within the Earth assume an iron-nickel-rich core (not unlike the iron-nickel meteorites) with a solid interior surrounded by a molten outer part. This combination could produce an electric dynamo that could sustain a magnetic field.

All these disciplines provide evidence of the Earth's formation in the early solar nebula. As more data are obtained, the picture of the Earth's origin becomes clearer and more refined.

Context

An understanding of the Earth's origin has many practical benefits. For example, the genesis of ore bodies is invaluable to prospecting for new resources. By knowing the products of various processes in the past, one is better able to predict human impact on present environments. Planetary engineering can use such information for the modification or preservation of conditions on the Earth, and maybe someday on other planets such as Mars. Meteorite size and shape studies were employed by space engineers in designing reentry vehicles and in studying their aerodynamic properties. Theories about material behavior in zero-gravity conditions similar to those in the solar nebula have led to experiments on manufacturing techniques in Earth orbit that are impossible to conduct on the Earth's surface.

Bruce D. Dod

Further Reading

Beatty, J. Kelly, Carolyn Collins Petersen, and Andrew Chaikin, eds. *The New Solar System*. 4th ed. Cambridge, Mass.: Sky, 1999. A general overview of the solar system and its components, this book has been organized around comparative planetology. A discussion of the various aspects of the genesis of the planets is scattered throughout. Draws heavily on the results of space exploration and on the interdisciplinary use of science to illustrate many concepts. Contains abundant illustrations, such as full-color photographs, artwork, graphs, and charts.

Brush, Stephen G. *Nebulous Earth: The Origin of the Solar System and the Core of the Earth from Laplace to Jeffreys*. New York: Cambridge University Press, 1996. An in-depth reference work detailing twentieth century theories on solar-system formation. Also discusses lunar origin theories.

Chaisson, Eric, and Steve McMillan. *Astronomy Today*. 6th ed. New York: Addison-Wesley, 2008. A well-written college-level textbook for introductory astronomy courses. Offers an entire chapter on the formation of planets.

Fraknoi, Andrew, David Morrison, and Sidney Wolff. *Voyages to the Stars and Galaxies*. Belmont, Calif.: Brooks/Cole-Thomson Learning, 2006. A well-written, thorough college textbook for introductory astronomy courses. Several sections deal with the origin of the solar system.

Freedman, Roger A., and William J. Kaufmann III. *Universe*. 8th ed. New York: W. H. Freeman, 2008. College-level introductory astronomy textbook, thorough and well written. Part of one chapter deals with the origin of the solar system, and part of another specifically addresses the development of the Earth.

Hartmann, William K. *Moons and Planets*. 5th ed. Belmont, Calif.: Thomson Brooks/Cole, 2005. A college textbook beyond the introductory level, with an approach based on comparative planetology. Offers much information on the origin of the solar system in general and the Earth in particular.

Horton, E., and John H. Jones, eds. *Origin of the Earth*. New York: Oxford University Press, 1990. A collection of articles written by experts in the field. Focuses on the study of the Earth and origins of the Earth-Moon system.

Hutchison, Robert. *The Search for Our Beginning*. New York: Oxford University Press, 1983. The author addresses the problem of determining the processes involved in the formation of the Earth and other solar-system bodies through the analysis of meteorites. Links astrophysics, geology, cosmochemistry, organic chemistry, and astronomy using meteoritics as the common ground. Provides historical perspectives along with space exploration results. Contains some fine illustrations, both in color and in black and white.

Morrison, David, and Tobias Owen. *The Planetary System*. 3d ed. San Francisco: Pearson/Addison-Wesley, 2003. Designed as a text for a college course in planetology, this book contains many references to the origins of the solar system and its individual components. Comparative planetology based on space exploration results, meteoritics, and other sources are utilized throughout the text to illustrate some of the evolutionary phases in the development of the planets and other solar-system objects. Extensively illustrated.

Ozima, Minoru. *The Earth: Its Birth and Growth*. Translated by J. F. Wakabayashi. New York: Cambridge University Press, 1981. Traces the genesis of the Earth and its growth while highlighting problems addressed by isotope geochemistry. The past 4.5 billion years are sketched in terms that are easy to comprehend. Alternative hypotheses and explanations are considered.

Schneider, Stephen E., and Thomas T. Arny. *Pathways to Astronomy*. 2d ed. New York: McGraw-Hill, 2008. A thorough college textbook for introductory astronomy courses, divided into many short sections on specific topics. An entire unit is devoted to the origin of the solar system.

Smart, William M. *The Origin of the Earth*. 2d ed. New York: Cambridge University Press, 1953. An older work useful for those interested in earlier hypotheses and theories of the Earth's origin.

Tarbuck, Edward J., and Frederick K. Lutgens. *Earth: An Introduction to Physical Geology*. Illustrated by Dennis Tasa. 9th ed. Upper Saddle River, N.J.: Pearson Prentice Hall, 2008. This college-level textbook for introductory geology courses is well written and fully illustrated. Contains a chapter on the origin and historical development of the Earth.

Wasson, John T. *Meteorites: Their Record of Early Solar-System History*. New York: W. H. Freeman, 1985. Written as a text for a course on solar-system genesis, this book includes topics on meteorite classification, properties, formation, and compositional evidence linking meteorite groups with individual planets. Describes how researchers use meteorites to determine conditions in the formative periods of Earth and other planets. Includes many graphs, charts, and illustrations.

EARTH'S ROTATION

Category: Earth

The rotation of the Earth results in the days and nights that provide the daily rhythm of life. Rotation causes the Earth to be flattened at the poles and to bulge at the equator. It also produces the Coriolis force, which influences the circulation of the atmosphere and oceans.

Overview

The spinning of the Earth on its polar axis is called rotation. The ancient Greeks considered the Earth to be a motionless body in the center of a geocentric (Earth-centered) universe. An exception was Heracleides (fourth century B.C.E.), who thought that the Earth did rotate. In general, the Greeks reasoned that, in their experience, if the Earth moved, they would feel some effects of it. Later, the work of Nicolaus Copernicus, Galileo Galilei, Johannes Kepler, and Sir Isaac Newton resulted in the paradigm shift to a heliocentric (Sun-centered) system in which the Earth was a planet simultaneously rotating on an axis while revolving around the Sun. The Earth rotates from west to east, thus making the Sun, Moon, planets, and stars appear to move from east to west across the sky. We commonly refer to the Sun, Moon, planets, and stars as rising and setting because it appears that they all are moving around the Earth, when in fact the Earth is rotating on its axis while revolving around the Sun.

The Earth's axis of rotation is inclined 23.5° from the perpendicular to the ecliptic (the plane of the Earth's orbit around the Sun); thus the axis of the Earth makes an angle of 66.5° to the plane of the ecliptic. The inclination causes the seasons (and the seasonal variation in the length of day and night) as the Earth orbits the Sun during the course of a year. When either the Northern or Southern Hemisphere of Earth is tilted toward the Sun, the period of daylight is longer and night is shorter in that hemisphere. The longer duration of daylight, coupled with the Sun's rays striking that hemisphere more nearly head-on, results in summer in that hemisphere. Conversely, when either hemisphere is tilted away from the Sun, the duration of daylight is shorter and night is longer, the Sun's rays strike that hemisphere more obliquely, and that hemisphere experiences winter.

Because of rotation, a point on the Earth's equator moves 1,674 kilometers per hour; the speed decreases to 1,450 kilometers per hour at 30° north or south latitude, and to 837 kilometers per hour at 60° north or south latitude. The Earth's rotation defines the unit of time called the "day." A day is defined as the interval of time between successive passages of a meridian, or line of longitude from the North to South Pole, under a reference object (for example, the Sun or a star). A day with reference to the Sun is called a solar day, and a day with reference to the stars is called a sidereal day. Because of the Earth's orbital motion around the Sun in one year, the Sun appears to move eastward relative to the stars approximately one degree per day. This makes the solar day approximately four minutes longer than the sidereal day, since the Earth must rotate a little bit farther to complete one rotation relative to the Sun as compared to the stars.

The Earth's orbit around the Sun is slightly elliptical, and the Earth's orbital speed varies with its distance from the Sun, being fastest when Earth is closest to the Sun (perihelion), around January 3, and slowest when Earth is farthest from the Sun (aphelion), around July 4. This means the Sun's apparent motion relative to the stars varies during the year, being greatest when the Earth's orbital speed is fastest at perihelion and smallest when the Earth's orbital speed is slowest at aphelion. Thus the length of the solar day as measured by a sundial (called the apparent solar day) varies slightly during the year. The length of the apparent solar day averaged over a year is called the mean solar day, and mean solar time is the basis for the 24-hour day (of 86,400 seconds) kept by clocks. (The sidereal day is 23 hours, 56 minutes, and 4.091 seconds long.)

Because of rotation, the Earth is flattened in the polar regions and bulges at the equator, thus making it slightly ellipsoidal, an oblate spheroid. The equatorial radius is 6,378 kilometers, while the polar radius is 6,357 kilometers. Thus a point on the equator is 21 kilometers farther from the center of the Earth than either pole is. Because the Earth is slightly flattened, the length of a degree of latitude changes from 109.92 kilometers at the equator to 111.04 kilometers at the poles. Also, because a point on the equator is farther from the center of the Earth and its rotational speed is faster, the effect of gravity is reduced there compared to other points on Earth. Thus an object at the equator weighs less, about 1 pound in 200, compared to the same object at either pole.

The Coriolis force or effect, named after a nineteenth century French engineer who studied this phenomenon, is caused by the Earth's rotation. It is an apparent force that affects free-moving bodies (such as wind, water, or missiles). For example, a fired projectile will veer to the right relative to the Earth's surface in the Northern Hemisphere, and to the left in the Southern Hemisphere. Precisely on the equator itself, free-moving objects are not deflected, but as they move north or south of the equator, the deflection becomes more pronounced.

The Coriolis effect is responsible for the global prevailing wind belts. Warm air near the equator rises and flows toward the poles. Cooled at higher altitude, the air descends around 30° north and south latitudes and spreads out both toward the equator and the poles. Air flowing toward the equator is deflected westward in both hemispheres, producing the easterly trade winds of the tropics. Air flowing toward the poles is deflected eastward in both hemispheres, producing the westerly winds of temperate latitudes. The Coriolis effect determines the direction wind blows around local high and low pressure systems in the atmosphere. Air moves outward from high pressure systems and inward toward low pressure systems. In the northern hemisphere, the moving air veers toward the right, setting up clockwise rotation around atmospheric highs and counterclockwise rotation around atmospheric lows. In the southern hemisphere, the moving air veers left, setting up counterclockwise rotation around highs and clockwise rotation around lows. This effect is especially noticeable in hurricanes (also called typhoons), which are regions of extremely low atmospheric pressure. The Coriolis force also influences ocean currents, which in turn affect the climate of coasts they flow along. The ocean currents of the Northern Hemisphere tend to flow clockwise, while those of the Southern Hemisphere flow counterclockwise. Witness the Gulf Stream of the north Atlantic and the Japanese (Alaskan) current of the north Pacific, always turning to the right, while the south Atlantic flow is to the left.

Overall, the rotation of the Earth is slowing down, and the length of the day is increasing by milliseconds per century. The decrease in rotational speed is primarily a result of the tidal friction caused by the gravitational pull of the Moon and to a lesser extent the Sun. Evidence for a lengthening day during geologic history comes from the study of fossils. Clams, corals, and some other marine invertebrates add a microscopically thin layer of new shell material each day, and the thickness varies seasonally throughout the year. Counting the daily growth lines in an annual set in well-preserved fossils yields the number of days in a year. Since the length of a year (the period of the Earth's orbit around the Sun) presumably has not changed, the length of a day in past geologic times can be determined. During the early Cambrian period (540 million years ago), there were 424 days in a year, and thus each day was about 20 hours, 40 minutes long. In the late Devonian period (365 million years ago), a year consisted of 410 days, each about 21 hours, 23 minutes long. At the beginning of the Permian period (290 million years ago), a year was down to 390 days, each about 22 hours, 29 minutes long.

As Earth's rotation slows in response to the Moon's tidal drag, the Earth's rotational angular momentum is transferred to the Moon, which increases the Moon's orbital angular momentum around the Earth, causing the Moon to move outward, away from Earth. This in turn increases the Moon's orbital period around Earth. This will continue until Earth's rotation on its axis is tidally synchronized with the Moon's revolution around Earth, both Earth and Moon keeping the same side facing each other. To conserve angular momentum, the Moon's distance and orbital period will increase to 549,000 kilometers (341,000 miles) and 46.7 of Earth's present days. Thus Earth's sidereal rotation period will be 46.7 days, and the mean solar day (time from noon to noon) will be 53.5 of Earth's present days. Since the length of the year will be unaffected, there will be only 6.8 solar days in a year.

Currently Earth's day is lengthening by an average of about 2×10^{-5} seconds each year, but the rate is quite erratic and sometimes even speeds up a bit. If this current average slowdown rate is extrapolated into the future, it will take 2×10^{11} years for Earth and Moon to become tidally locked. Alternatively, laser ranging (using retroreflectors left on the Moon's surface by the Apollo Moon landings) shows that the Moon currently is moving away from Earth at 3.8 centimeters per year. Extrapolating this rate into the future, it will take 4×10^9 years for the Moon to reach its final distance. The disagreement of these two time estimates indicates that the rates of slowdown of Earth's rotation and the increase in the Moon's distance

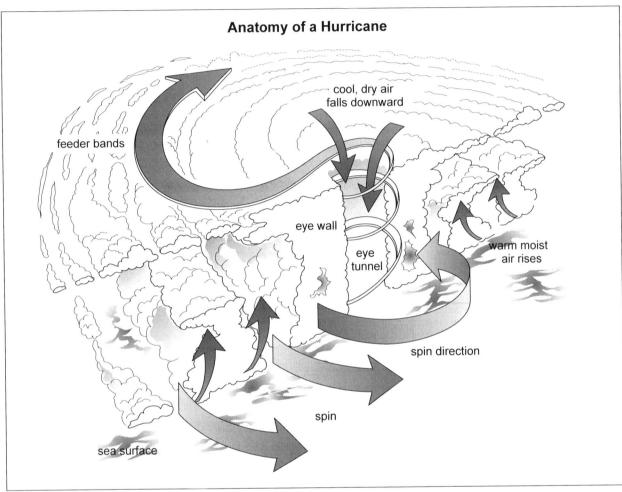

Anatomy of a Hurricane

feeder bands

cool, dry air
falls downward

eye wall

eye
tunnel

warm moist
air rises

spin direction

spin

sea surface

The Coriolis force, which influences the circulation of the atmosphere and oceans, is a result of Earth's rotation and determines the direction wind blows around local high and low pressure systems in the atmosphere. The effect is especially noticeable in hurricanes, regions of extremely low atmospheric pressure.

probably will not remain constant, and the time to achieve Earth-Moon tidal lock is probably at least billions of years. However, before this can occur, the Sun probably will expand and become a red giant first.

Numerous detailed studies show that, superimposed on the systematic long-term slowdown, Earth's rotation has numerous small random changes. The reasons for these variations include transfer of angular momentum between different parts of the Earth's interior; transfer of angular momentum between the atmosphere, the oceans, and the Earth's surface; movement of air masses and changes in wind patterns; growth or shrinkage of polar ice caps; volcanic activity; earthquakes; and plate tectonic movements. The timescales for the various effects

on the Earth's rotation vary, from short-term to long-term, and from systematic seasonal to erratic. It is evident from many studies that the Earth's rotational speed has varied throughout geologic time and continues to change on a daily, weekly, monthly, seasonal, yearly, and even longer-term basis.

Currently the Earth's axis of rotation points toward Polaris, Earth's present North Star. However, the Earth's axis slowly changes direction in space in a process called precession. Recall the Earth has an equatorial bulge, and it is tilted about 23.5° from the ecliptic, the plane of the Earth's orbit around the Sun. The torque exerted by the gravitational pull of the Moon and the Sun on the Earth's equatorial bulge trying to make it line up with the ecliptic plane causes the Earth's axis to slowly precess, like the axis of a tilted spinning toy top. This precession causes the Earth's axis to trace out in space a double cone (two cones joined at their vertices) with a vertex angle of 47° (twice the 23.5° axial tilt). The Earth's axis slowly shifts direction about 50 arc seconds per year, and it takes about 26,000 years for a complete precession cycle. As the axis points to different parts of the sky in response to precession, stars other than Polaris have served and will serve as north stars. Also, in about 13,000 years (half the precessional cycle), the constellations seen during specific months on Earth will be "shifted" by six months, so that those constellations now seen during June (for example) will be seen during December, and so on. Consequently, the astronomical coordinate system of right ascension and declination slowly and systematically changes during the precessional cycle. Catalogs listing those coordinates for stars, nebulae, galaxies, and other celestial bodies must specify the epoch (year) for which the listed coordinates are rigorously correct. To point a telescope at some desired object some other year requires calculating precessional corrections to the listed coordinates.

Superimposed upon the precessional motion are two other motions. One of these is a small oscillating motion called nutation, which has a semiamplitude of 9.2 seconds of arc and a period of 18.6 years. This motion is associated with the periodic variation in the orientation of the Moon's orbital plane around the Earth with the Earth's orbital plane around the Sun. The other motion, called Chandler's wobble, has two oscillations. One of the oscillations, the Chandler component, with a period of twelve months, is a result of meteorological effects associated with seasonal changes in air masses. The second oscillation of the Chandler wobble, the 14.2-month component, is caused by shifts in the Earth's interior mass. Thus the changing direction of the Earth's rotational axis is not smooth but "wiggly" or "wobbly."

Methods of Study

Sundials were first used to mark the passage of the apparent solar day, as the shadow of the gnomon (stick) moved across the face of the dial. In 1671, the French astronomer Jean Richer made time measurements with a pendulum clock both in Paris (49° north) and in Cayenne, French Guiana (5° north) and compared the two. In French Guiana, the clock "lost" 2.5 minutes per day compared to Paris. He attributed this loss to a decrease in effective gravitational pull toward the equator due to the Earth's

rotation; the practical consequence was that pendulum clocks needed to have the length of their pendula adjusted according to latitude to be able to keep accurate time.

In 1851, the French physicist Jean-Bernard-Léon Foucault hung a 25-kilogram iron ball with a 60-meter-long wire from the dome of the Panthéon in Paris, with a pin at the bottom of the ball to make marks in a smooth layer of sand underneath. After only a few minutes, the tracings in the sand showed that the plane of the ball's swing slowly rotated clockwise as seen from above. Foucault explained this as a demonstration of the Earth's rotation, which moved the attachment point on the dome and the sand on the floor, while the pendulum tried to maintain the plane of its swing in the same direction. In the 1950's, atomic clocks began to be used to measure time accurately over long periods. When time kept by these clocks was compared to time determined by the rotation of the Earth, small variations in the Earth's rotation were found.

Newer techniques used to determine length of day and polar motion involve the use of satellites and lasers. One method, called lunar laser ranging (LLR), involves the emission of light pulses from a laser on Earth to reflectors left on the Moon by Apollo and Soviet spacecraft. The returning pulses of light are received by a telescope. The total travel time is calculated to determine the Earth-to-Moon distance. By observing the time the Moon takes to cross a meridian during successive passages, this method has provided very good length-of-day measurements. Another technique involves the use of the Laser Geodynamics Satellite (Lageos). This satellite is covered by prisms that reflect light from pulsed lasers on Earth. Again, the returned beam is received by a telescope and the round-trip travel time is used to infer the one-way distance from the Earth to the satellite. This method, which includes a network of stations on Earth, can provide insight into yearly movement of crustal plates, which is believed to cause variations in the Earth's rotation. A very accurate technique known as very-long baseline interferometry (VLBI) is also being used to plot continental drift as well as variations in Earth's rotation and the position of the poles. In this method, radio signals from space (typically from quasars) are received by two radio antennas and are tape-recorded. The tapes are compared, and the difference between the arrival times of the signals at the two radio antennas is used to calculate the distance between the two. If the distance between the two antennas has changed, the crustal plates have moved.

Context

The spinning of the Earth on its polar axis once every twenty-four hours is very much a part of the daily rhythm of life. Among the primary ways Earth's rotation is felt by and governs life are its impact on our day, on gravitational pull, and on the atmosphere. Earth's rotation gives us a daily time reference by the passage of days and nights. The spinning of the Earth on its polar axis causes the Earth to bulge at the equator and to be flattened in the polar areas. Because of this phenomenon, the distance to the center of the Earth varies with latitude, and as a result the effective gravitational pull on objects on the Earth's surface also varies—objects weigh slightly less at the equator than in the polar regions of the world. The Coriolis effect, an apparent force caused by the rotation of the Earth, causes free-moving bodies to be deflected to the right in the Northern Hemisphere and to the left in the Southern Hemisphere. It governs the direction of winds as they flow in or out of pressure systems, establishing the wind belts of the world. The Coriolis force also affects the flow of ocean currents, and these patterns help to alter climates along the coasts of continents.

Roberto Garza

Further Reading

Bostrom, Robert C. *Tectonic Consequences of the Earth's Rotation*. New York: Oxford University Press, 2000. Reviews scientific data looking for a link between geotectonics and the Earth's rotation. The author presents a better theory to explain tidal Earth. For the more advanced reader or undergraduate.

Gould, S. G. "Time's Vastness." *Natural History* 88 (April, 1979): 18. This article summarizes the reasons for the slowing down of the Earth's rotation. It discusses the use of corals as a proof that the length of the day is increasing and that the number of days in a year is decreasing. Suitable for high-school-level readers.

Lambeck, Kurt. *The Earth's Variable Rotation: Geophysical Causes and Consequences*. New York: Cambridge University Press, 2005. Focuses on the irregular rotation of the Earth and gives detailed analysis of the various reasons for it. Covers the interdisciplinary fields of solid Earth physics, oceanography, meteorology, and magnetohydrodynamics. Technical.

McDonald, G. E. "The Coriolis Effect." *Scientific American* 186 (1952): 72. The article takes a nontechnical approach to the study of how objects move on the Earth as a result of the Coriolis effect. Suitable for high school readers.

Markowitz, W. "Polar Motion: History and Recent Results." *Sky and Telescope* 52 (August, 1976): 99. This article reviews studies of polar motion, looking at how the Earth's rotation and precessional motions are affected by various forces.

Mulholland, J. D. "The Chandler Wobble." *Natural History* 89 (April, 1980): 134. Discusses how small movements affecting the Earth's axis may be associated with other terrestrial phenomena. Suitable for high school readers.

Munk, W. H., and G. J. F. MacDonald. *The Rotation of Earth: A Geophysical Discussion*. New York: Cambridge University Press, 1960, Dated, but valuable for its detailed analytical treatment of the physics of Earth's rotation. Includes discussion of the small fluctuations in rotation as a result of redistribution of angular momentum, thought to be caused by dynamics in the fluid outer core. Designed for professional geophysicists.

Rosenburg, G. D., and S. K. Runcorn, eds. *Growth Rhythms and the History of the Earth's Rotation*. New York: John Wiley & Sons, 1975. A compilation of studies that can serve as an introduction to the methods of determining the history of the Earth's rotation. The text is suitable for college-level readers not intimidated by technical language. Each study includes a bibliography, and the book is carefully indexed by author, taxonomy, and subject.

Smylie, D. E., and L. Mansinha. "The Rotation of the Earth." *Scientific American* 225 (December, 1971): 80. This article analyzes measurements indicating that the Earth's wobble may be due to earthquakes. It is a well-illustrated article that can be understood by high school readers.

Stephenson, F. Richard. *Historical Eclipses and Earth's Rotation*. New York: Cambridge University Press, 2008. Investigates the history of the Earth's rotation by studying eclipses throughout ancient and medieval times. Shows how tides cannot be solely responsible for the lengthening of the day.

Tarbuck, Edward J., and Frederick K. Lutgens. *Earth: An Introduction to Physical Geology*. Illustrated by Dennis Tasa. 9th ed. Upper Saddle River, N.J.: Pearson Prentice Hall, 2008. This college-level textbook for introductory geology courses gives a very clear explanation of the tidal slowdown of the Earth's rotation and specific details on the number of days in the year in past geologic times based on fossil evidence.

EARTH'S SHAPE

Category: Earth

It has been known for centuries that Earth is not a perfect sphere. The diameter of the planet is greater at the equator than it is from pole to pole. This oblateness is the result of Earth's daily rotation on its axis. Even smaller irregularities in shape have been measured by Earth-orbiting satellites.

Overview

The discovery that Earth is not a perfect sphere dates to the seventeenth century, when measurements of the distance corresponding to one degree of latitude were found to increase systematically from the equator toward both poles. Because of Earth's oblateness, one degree of latitude has a length of 110.6 kilometers at the equator and 111.7 kilometers at the poles.

Besides rotation, other forces imposed upon Earth affect the shape of the planet, a prime example being Earth tides. The oceans of the world generally have two tidal bulges, or regions where the ocean surface is relatively high, caused primarily by the gravitational attraction of the Moon and to a lesser extent by the Sun. When the Sun, Moon, and Earth all are on a straight line (the Sun and Moon either on the same side or opposite sides of the Earth), the oceans display the highest high tides and lowest low tides (spring tides), as the tidal effects of the Sun and Moon reinforce each other. When the line from the Earth to the Moon makes a right angle to the line from the Earth to the Sun, the tidal effects of the Sun and Moon partially cancel out, and the high tides are not very high and the low tides are not very low (neap tides). Not only does the water of the oceans rise and fall because of tides, but so too the surface of the "solid" Earth rises and falls very slightly due to the same tidal forces imposed by the Moon and the Sun. This periodically varying distortion is so slight as to render accurate measurements of it quite difficult.

A view of Earth from satellite distance in space would, to the naked eye, suggest that the planet is a perfect sphere, yet measurements reveal that it is not. All planets (including Earth), along with the larger satellites and

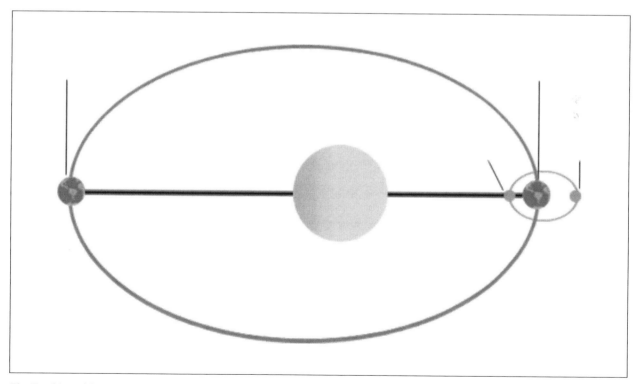

The Earth's and Moon's elliptical orbits—Earth's around the Sun, and the Moon around Earth—affect their tides and hence their shapes. (NOAA)

asteroids, are essentially spheroidal, while smaller objects are not. Sufficiently small solid objects (such as books, boulders, and bones) can maintain any arbitrary shape because of the strength of the material of which they are composed. However, sufficiently large objects (even if solid) have internal forces due to their self-gravity that are strong enough to overwhelm the strength of whatever material composes them, and they pull themselves into spherical shapes. The critical "threshold" size depends on the density and strength of the solid material, but for the three types of solids (ices, silicate minerals, and metals, mainly iron) most common in the solar system, the threshold is approximately the same: on the order of a few hundred kilometers.

Rotation (spinning on an axis) makes a large object depart from being spherical. The Earth's rotational period, measured in a quasi-inertial frame based on distant stars and galaxies, is 23 hours, 56 minutes, and 4 seconds. As a result of this rotation, a point on the Earth's equator moves with a speed of 1,674 kilometers per hour, a point at 30° north or south latitude moves 1,450 kilometers per hour, and a point at 60° north or south latitude moves 837 kilometers per hour. The increase in rotational speed toward the equator makes the equator bulge outward, transforming a spherical shape into an oblate spheroid. The Earth's equatorial diameter is 12,756 kilometers, while its polar diameter is 12,714 kilometers. Its oblateness is defined as the difference in diameters divided by the equatorial diameter, which gives a value of 0.00336 or about one part in 298.

A comparison of the planets in the solar system shows that rotation plays a dominant role in determining oblateness. Mercury and Venus, each with very slow rotation, have no discernible oblateness and are essentially spherical. Jupiter and Saturn rotate the fastest, and both are noticeably oblate as seen in telescopes and spacecraft images. However, there are other factors. For example, Saturn rotates slightly more slowly than Jupiter, and Mars rotates slightly more slowly than Earth; yet Saturn is more oblate than Jupiter (about one part in 10 compared to one part in 15), and Mars is more oblate than Earth (about one part in 200 compared to one part in 300). These "discrepancies" probably are due to the distribution of mass throughout the planet and the rigidity of the material composing the different parts of the planet.

Methods of Study

Measurements of small departures from an oblate spheroid shape became possible with the advent of the space age and the development of twentieth century instrumentation. Perturbations in the orbits of Earth-orbiting satellites show that the Earth's mass is not distributed as it would be were it a simple oblate spheroid. Some of the earliest data indicated that the Earth is slightly pear-shaped, with small bulges of up to about 100 meters near the North Pole and in a band south of the equator. More recently, satellite-mounted radar altimeters have been able to map continuously the topography of the ocean surface to an accuracy of a few centimeters by bouncing radar signals off the water. With wave crests and troughs averaged out, ocean surfaces show small but significant deviations from a smooth oblate spheroid that reflect the topography of the seafloor underneath. Major seamounts and suboceanic ridges are clearly marked by regions of higher ocean surface above. Likewise, the major deep-sea trenches, as are common around the rim of the Pacific Ocean, are marked by troughs in the ocean surface above. This phenomenon is due to small variations in the acceleration of gravity. In the case of a seamount or ridge, there is a concentration of mass (the rock composing the feature), so gravity there is a bit stronger and attracts more water over it. In contrast, there is a deficit of mass in a trench, so gravity there is slightly weaker and attracts less water over it.

Context

The passing of geologic time has brought changes in the phenomena that control Earth's shape. The distance from the Earth to the Moon was less than at present, and Earth rotated faster in the past. These two differences would have produced larger and stronger tides, and a larger equatorial bulge. It is interesting to speculate as to what effects those changes might have had on ancient dynamic processes. Today, Earth's inhabitants suffer little effect from the planet's distortion. It cannot be observed with the naked eye, and it does not appear to play a role in weather patterns and climate.

However, one practical, though small, consequence of the Earth's shape is the variation of the effective gravitational acceleration with latitude. Due to Earth's equatorial bulge, which is the result of Earth's daily rotation on its axis, the effective gravitational acceleration, and hence the weight of objects, is slightly less at the equator than at either pole. Because of the equatorial bulge, objects at sea level at the equator are about 21 kilometers farther from the center of Earth than if they were at sea level at either pole. This alone reduces the gravitational acceleration at the equator by about 0.15 percent compared to the value at either pole. The effective gravitational acceleration at the equator is further reduced directly by Earth's rotation, since some of the gravitational acceleration that would otherwise exist is used to provide the centripetal acceleration needed to make objects follow the curved paths that keep them in contact with Earth's surface as Earth rotates. This reduces the effective gravity at the equator by about 0.35 percent. Combining the two effects, the gravitational acceleration varies from 9.832 meters per second squared at either pole to 9.780 meters per second squared at the equator, or about 0.5 percent. As a result, weight at the equator is reduced by about 0.5 percent compared to weight at either pole, so an object weighing 200 pounds at either pole weighs about 1 pound less at the equator.

John W. Foster

Further Reading

Greenberg, John L. *The Problem of the Earth's Shape from Newton to Clairaut*. New York: Cambridge University Press, 1995. Covers the early studies to determine the shape of the Earth by various scientists, including Isaac Newton. Explains their influence on Alexis Claude Clairaut, who confirmed Newton's belief that the Earth is flattened at the poles.

Ince, Martin. *The Rough Guide to the Earth 1*. New York: Rough Guides, 2007. A handy reference on most aspects of Earth science. Includes several diagrams and pictures to help explain material. For beginners: nontechnical and easy to read.

James, David E., ed. *The Encyclopedia of Solid Earth Geophysics*. New York: Van Nostrand Reinhold, 1989. A complete reference work on solid-Earth geophysics. Includes more than 150 articles by top scientists. Also covers topics such as geology, seismology, and gravimetry, among others. For advanced readers: detailed and technical.

King-Hele, D. "The Shape of the Earth." *Scientific American* 217 (October, 1967): 17. This article begins with the historical views of the shape of the Earth. It then discusses, with a good set of illustrations, how satellites have helped scientists to learn more about the shape of the Earth. The nontechnical approach makes the article suitable for high school and general readers.

Melchior, Paul. *The Earth Tides*. Oxford, England: Pergamon Press, 1966. A sophisticated treatment of the physical phenomenon of small distortions of Earth resulting from gravitational forces imposed by the Moon and the Sun.

Stacey, Frank D. *Physics of Earth*. 2d ed. New York: John Wiley & Sons, 1977. A reference volume on solid-Earth geophysics, including radioactivity, rotation, gravity, seismicity, geothermics, magnetics, and tectonics. Provides detailed numerical tabulations on dimensions, properties, and unit conversions.

EARTH'S STRUCTURE

Category: Earth

Processes that occur in the interior of the Earth have profound effects upon the surface of the Earth and its human population. The results of such processes include earthquakes, volcanic activity, and the shielding of life-forms from solar radiation.

Overview

A simple demonstration that the Earth's interior is different from its surface is to compare the Earth's average density, calculated by dividing its mass by its volume, with the density of typical rocks from the surface. The average density is about 5.5 grams per cubic centimeter, while the density of typical surface rocks is between about 2.7 and 3.0 grams per cubic centimeter. This means that part of the interior must be composed of much denser material than surface rocks.

Evidence that the interior is differentiated into "layers" of various thicknesses, compositions, and mechanical properties comes primarily from analyzing the seismic waves produced by earthquakes that travel through the Earth. The thinnest layer is the outermost one known as the crust. The crust varies in thickness from about 5 kilometers under parts of the ocean basins up to about 70 kilometers under the highest mountain ranges of the continents. The crust is composed of a number of different rock types, but there are systematic differences between the crust of continents and that of ocean basins; continental crust is generally granitic (similar to granite), while oceanic crust is generally basaltic (similar to basalt). Both granite and basalt are igneous rocks, meaning that they cooled and hardened from hot molten material, and both are composed of silicate minerals.

However, the silicate minerals in basalt (such as pyroxene, olivine, and calcium-rich plagioclase feldspar) are comparatively rich in iron, magnesium, and calcium, giving them a generally dark color and slightly greater density (about 3.0 grams per cubic centimeter), while the silicate minerals in granite (such as quartz, potassium feldspar, and various micas) are poorer in iron, magnesium, and calcium, making them generally lighter in color and slightly lower in density (about 2.7 grams per cubic centimeter). Note that this distinction does not hold completely: Basalt and similar rocks can be found on continents, and sediments weathered from granite and similar rocks can be found in ocean basins.

The base of the crust is marked by a boundary known as the Mohorovicic discontinuity, or Moho. It represents a change in density of the rock above and below it. Rocks just below the Moho are slightly denser, about 3.3 grams per cubic centimeter, than either continental or oceanic crustal rocks. The rocks below the Moho probably are peridotite. Peridotite, composed mostly of olivine and pyroxene, is similar to basalt, but it is richer still in iron and magnesium. Peridotite is thought to represent the general composition of the layer underlying the crust, called the mantle. The mantle comprises the bulk of the Earth, representing about 80 percent by volume.

In the upper mantle, at depths starting about 100 kilometers beneath the surface and extending down to about 410 kilometers, is a zone of less rigid and more plastic, perhaps even partially melted, material called the asthenosphere. The crust and the part of the mantle above the asthenosphere, acting as a rigid unit, are known collectively as the lithosphere. The change to plastic behavior in the asthenosphere occurs because temperatures there are close to the melting point of peridotite. Although temperature continues to increase below the asthenosphere, the greater pressures at greater depths are high enough to keep the rock from melting.

The asthenosphere is thought to play an important role in movements of the lithosphere above. According to the theory of plate tectonics, the lithosphere is divided into a number of plates about 100 kilometers thick that are in constant motion at speeds of up to several centimeters per year, driven by hot convection currents of material moving slowly in the plastic asthenosphere. The hot material rises along divergent plate boundaries marked at the surface by the volcanic ridge-rift system that extends through the ocean basins around the globe. The slowly moving convection currents in the asthenosphere then move laterally away from the ridge-rifts, carrying the lithospheric plates above away from the ridge-rifts. As it moves laterally, the asthenosphere cools, becoming denser and sinking back downward. The sites where the convection currents sink are places where lithospheric plates with ocean crust on top dive into the mantle in a process called subduction.

At these sites, marked at the surface by trenches in the ocean basin floor, crustal rocks may be carried into the upper mantle to depths as great as 700 kilometers. Below this level, the rock may simply be too dense for the lithospheric plates to penetrate.

There are two lower boundaries within the mantle. At 410 and 660 kilometers below the surface, abrupt increases in density occur. Although one might suspect a change in composition to account for the jump in density, laboratory studies of rocks under pressure suggest an alternative explanation. The primary mineral in peridotite is olivine. The pressures at 410 kilometers and again at 660 kilometers collapse the crystalline structure and produce denser minerals with the same iron and magnesium silicate composition. At the pressure existing at 410 kilometers, olivine converts to the denser mineral called spinel, and at the even higher pressure at 660 kilometers, both spinel and pyroxene collapse to yet a denser mineral known as perovskite. Thus the changes occurring in the mantle to produce the asthenosphere and the discontinuities at 410 and 660 kilometers are not changes in composition but instead changes in physical properties caused by temperature and pressure. The density increases from about 3.3 grams per cubic centimeter at the top of the mantle to about 5.6 grams per cubic centimeter at its base.

The next layer beneath the mantle is the outer core. This layer begins at a depth of about 2,900 kilometers beneath the surface and continues to a depth of 5,100 kilometers. There is a large density increase across the core-mantle boundary, from 5.6 grams per cubic centimeter at the base of the mantle to about 10 grams per cubic centimeter at the top of the core. Iron is the only reasonably abundant element that would have the required density at the tremendous pressure of millions of atmospheres at these depths. However, pure iron would give too high a density, so iron mixed with about 15 percent nickel, sulfur, silicon, and possibly oxygen and even hydrogen has been suggested. At the pressures and temperatures that must exist in the outer core, iron alloys would be in a molten state. Complex currents of metallic iron alloy, generated in the fluid outer core by convection and the Earth's rotation, give rise to the Earth's main magnetic field through a geodynamo process.

The core-mantle boundary represents a composition change from the silicate minerals of the lower mantle to the metals of the core. The boundary is a sharp one, but whether it is smooth and spherical in shape or irregular with "hills" or "peaks" on its surface is not known. There is some evidence from seismology that the lower mantle within 100 kilometers of the core boundary is a transition zone with a change of properties. It may consist of a mix of mantle and core material that is less rigid than the mantle rocks above it.

The innermost layer of the Earth's interior is the inner core. This region has a radius of about 1,300 kilometers where there is a boundary with the outer core. Increasing pressure at these depths requires that the iron of the inner core is solid. It is thought the solid inner core continues to grow in size as iron in the molten outer core crystallizes as Earth slowly cools. Because the solid inner core is separated from the mantle by the molten outer core, it can rotate independently. Seismic studies suggest that the inner core rotates slightly faster than the mantle and crust.

Methods of Study

Much of what is known about the structure of the Earth's interior comes from the analysis of seismic waves generated by earthquakes or by explosives detonated at or just below the surface. After passing through the Earth, the wave vibrations are recorded on seismographs located all around the world, revealing information about the part of the interior they traveled through.

The seismic waves that pass through the interior are called body waves, because they propagate through the body of the Earth and not along the surface. Body waves are of two varieties: primary (or P) waves, and secondary (or S waves). P waves are the same as acoustic or sound waves. They cause the material they traverse to move back and forth in the direction of wave travel, alternately stretching and compressing it. Like ordinary sound waves, P waves can travel through any sort of material—solid, liquid, or gas. S waves are transverse waves, which move material along the wave path from side to side. Consequently they can travel only through rigid, that is, solid, material; S waves cannot travel through liquids or gases.

Both P waves and S waves cross the asthenosphere of the upper mantle, but with reduced speed (that is why the upper mantle is sometimes called the low-velocity zone), suggesting lower rigidity but not a liquid state, since S waves do propagate through it. Therefore, it seems that the asthenosphere is a solid but plastic region, able to ooze and flow very slowly.

At the core-mantle boundary, P waves abruptly slow down by almost a factor of 2 as they enter the outer core, and S waves disappear, indicating the material of the outer core has no rigidity. Since gases cannot exist at the

conditions of the outer core, the material of this region must be a liquid.

Other locations in the interior are marked by increased speeds for both P waves and S waves. There is a sharp increase in speed at the Mohorovicic discontinuity at the base of the crust. The P-wave speed jumps from about 6 kilometers per second above the Moho to around 8 kilometers per second below it, while the S-wave speed jumps from about 4 to 5 kilometers per second.

Below the asthenosphere, both P-wave and S-wave speeds gradually increase to a depth of about 410 kilometers, at which point both sharply increase. Laboratory studies indicate that the increase in density when olivine collapses to form spinel accounts for the increased speeds at this depth. At a depth of around 660 kilometers, a second abrupt increase in both wave speeds occurs, here caused by a second collapse to produce the yet denser mineral perovskite. The speeds for this part of the mantle match those obtained in the laboratories from waves passing through perovskite samples placed under the kinds of pressures found at 660 kilometers. A final increase in speed is observed when P waves pass the outer to inner core boundary. This is due to a phase transition from liquid iron in the outer core to solid iron in the inner core. Such a phase transition is supported by the probable reappearance of S waves in the inner core.

Another way in which the existence of structural boundaries within the Earth can be shown from an analysis of seismic waves is to examine the way the waves reflect and/or refract when they encounter the boundaries. What the waves do depends on the angle at which they approach the boundary, as well as on the properties of materials on both sides of the boundary. P waves are reflected off the Moho, the core-mantle boundary, and the inner-outer core boundary, providing clear evidence that there are sharp boundaries between the crust and mantle, the mantle and outer core, and the outer and inner cores. Waves have also been detected reflecting off the 660-kilometer discontinuity.

Refraction or bending of waves yields further evidence. As P waves cross the mantle-core boundary, they are refracted toward the center of the Earth because their speed is less in the molten outer core. This deflection of P waves passing through the outer core leaves a gap stretching around the Earth in the form of a band extending from 100° to 140° from the epicenter of the earthquake. This gap is known as the P-wave shadow zone because no P waves reach the surface in this band.

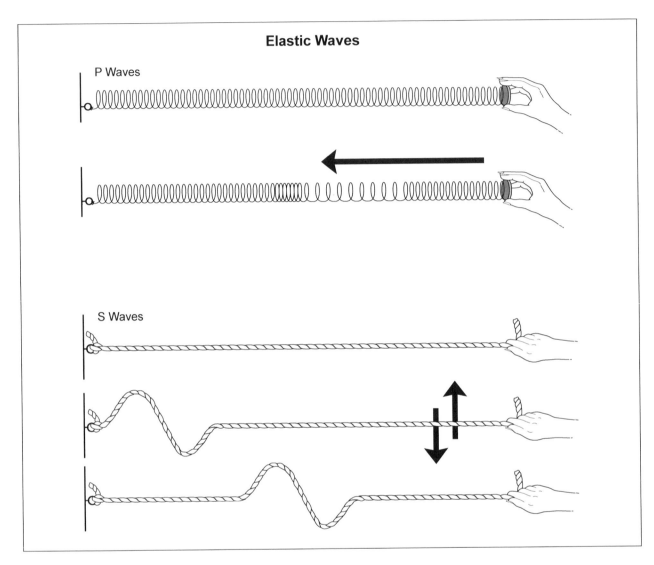

Advances in computer science have allowed the identification of even subtler details about the Earth's interior. Computerized tomography is a technique used in medicine, in which X rays from all directions are analyzed in a computer to give a three-dimensional picture of the internal organs of the human body. Seismic tomography is an analogous approach that uses seismic waves that travel from earthquakes to seismographs around the world to map the Earth's interior. This technique includes both P and S body waves traveling through the interior as well as surface waves traveling along the surface of the Earth. By looking at the travel times of the different waves, scientists are able to compare speeds along different paths. Such an approach has resulted in maps of slow and fast regions of the mantle that probably represent less rigid (warmer) and more rigid (cooler) regions.

Context

The interior of the Earth has profound effects on the surface. The interior acts as a complex heat engine that is the driving force behind plate tectonics, resulting in the formation and evolution of oceanic and continental crust. In the process, earthquakes and volcanic activity occur that create hazards for the human population on the Earth's surface. Complete acceptance of the plate tectonic theory could not have occurred without the discovery of the asthenosphere, which makes the movement of the lithospheric plates plausible.

It is fortunate that the Earth has a magnetic field. Without it, the age of discovery and exploration would not have been possible, for navigation by magnetic compasses allowed voyages across uncharted oceans. It is now theorized that the Earth's magnetic field is generated in the molten metallic outer core by the geodynamo process. An important effect of the core-generated magnetic field is the changes it has undergone through time. In particular, at rather irregular intervals of geologic time, the magnetic poles reverse polarity; during such reversals, the magnetic field decreases in strength. Since the magnetic field shields life-forms on the Earth's surface from charged particles emitted by the Sun, there is some concern that such field weakenings during polar reversals could result in more cancers and genetic mutations. Some scientists suspect that polar reversals might be at least partly responsible for some of the mass extinctions that have occurred in the geologic past, as well as periods of rapid evolution. Thus, the surface of the Earth as well as the life-forms on it depend upon and are strongly affected by processes occurring within the Earth's interior.

David S. Brumbaugh

Further Reading

Bolt, Bruce A. "Fine Structure of the Earth's Interior." In *Planet Earth*. San Francisco: W. H. Freeman, 1974. An extremely well illustrated review of how seismic waves have been used to discover and define the various layers of the Earth's interior. Written at a general-interest college level. Little background or expertise in mathematics is required.

_____. *Inside the Earth: Evidence from Earthquakes*. San Francisco: W. H. Freeman, 1982. This book is written for undergraduate college students in physics and the Earth sciences and for nonspecialists interested in a more detailed summary of knowledge of the Earth's interior. The text is relatively free of mathematics and is clearly and well illustrated. It offers a concise and readable treatment of the use of seismic waves to discover and interpret the Earth's interior.

Fowler, G. C. *The Inaccessible Earth: An Integrated View to Its Structure and Composition*. 2d ed. New York: Chapman and Hall, 1993. A commonly used introductory geophysics text. Readers should have knowledge of basic calculus. Includes a series of questions and problems.

Tarbuck, Edward J., and Frederick K. Lutgens. *Earth: An Introduction to Physical Geology*. Illustrated by Dennis Tasa. 9th ed. Upper Saddle River, N.J.: Pearson Prentice Hall, 2008. This college-level textbook for introductory geology courses is well written and illustrated. There is a very good chapter on the Earth's interior, as well as several chapters on volcanoes, earthquakes, and plate tectonics that describe their relationships with the interior structure and processes.

Van der Pluijm, Ben, and Stephen Marshak. *Earth's Structure*. 2d ed. New York: W. W. Norton, 2003. An introductory text on structural geology and tectonics. Designed for undergraduate students.

Vogel, Shawna. *Naked Earth: The New Geophysics*. New York: Plume, 1996. Covers the field of geophysics from 1960 with the theory of plate tectonics through the mid-1990's. Written with little technical jargon and many examples.

EARTH-SUN RELATIONS

Categories: Earth; Planets and Planetology; The Sun

The fundamental Earth-Sun relationship, from which all others derive, is that the Earth rotates on its axis as it revolves around the Sun. The relationships between the Earth and the Sun determine the Earth's "heat budget" and control life on the planet. Earth motions also produce noticeable periodic changes in the apparent path of the Sun across the sky, perhaps most obvious in the seasonal change in directions of sunrise and sunset, and the length of time the Sun is above the horizon.

Overview

Earth-Sun relations are the dominant controls of life on Earth. The Sun is a star, and its electromagnetic radiation warms the Earth and supplies the energy that supports life on the planet. Earth-Sun relations determine the amount, duration, and distribution of solar radiation that is received by Earth. The Earth's rotation on its axis produces day and night, and its revolution around the Sun and the tilt of its rotational axis result in the seasons; these processes serve to distribute solar radiation over the Earth. Earth's atmosphere and oceans influence the reflection, absorption, and transfer of solar energy. The result of these interacting phenomena is a "heat budget" on Earth that is hospitable to life.

The Sun radiates electromagnetic energy from every part of its spherical surface. Earth, 150 million kilometers away, intercepts only a minute portion of the Sun's radiation, about one two-billionth. The small amount of the Sun's energy that strikes Earth is Earth's energizer. It sustains life on Earth and drives weather systems and oceanic circulation. Solar energy from the past has been preserved in the form of fossil fuels—coal, petroleum, and natural gas.

Perhaps the most remarkable aspect of Earth—remarkable because it is rare in our solar system—is its relatively narrow range of moderate temperatures. The adjectives "hot" and "cold" are frequently used in describing our weather. In relation to the temperatures that are found elsewhere in the solar system, Earth is always moderate, and the words "hot" and "cold" better describe conditions on the other planets. The mean temperature of Earth is about 15° Celsius (59° Fahrenheit, or 288 kelvins); the absolute extremes recorded anywhere on Earth are 58° Celsius (136° Fahrenheit, or 331 kelvins) in North Africa and −89° Celsius (-128° Fahrenheit, or 184 kelvins) in Antarctica. Few inhabitants of Earth will ever experience a temperature range of much more than 60° or 70° Celsius in a lifetime. Compare these temperatures with those of Earth's nearest neighbor, the Moon, where temperatures range from about 120 to -170° Celsius (393 to 103 kelvins) between the day- and nightsides. Earth's so-called sister planet, Venus, has a surface temperature of about 450° Celsius (723 kelvins). The outer planets of the solar system experience a permanent deep freeze, below -100° Celsius (137 kelvins).

The best demonstration of the moderate nature of Earth's temperature is the presence of the world's oceans. Water can exist in the liquid state only in the narrow temperature range of 0 to 100° Celsius (273 to 373 kelvins) at Earth's surface atmospheric pressure. Yet almost 98 percent of Earth's water remains in the liquid state. The polar ice caps contain 2 percent, and a minute portion is water vapor in the atmosphere at any time. Currently 71 percent of Earth's surface is covered by oceans of liquid water, and it has had large oceans for much of its existence as a planet.

The factors that cause Earth to experience such moderate temperatures are complex and interrelated. The Sun is the source of the energy, yet being the right distance from the Sun cannot be the sole cause of Earth's moderate temperature—witness the Moon. Rather, the explanation has to do with Earth's atmosphere, its oceans, and its motions relative to the Sun.

Earth's atmosphere moderates the planet's temperature during both daylight and darkness. During daylight, the atmosphere blocks excessive amounts of solar radiation from reaching Earth's surface and thus prevents overheating. During darkness, the atmosphere retards the escape of heat in the form of long-wave infrared energy back into space and thus prevents excessive overnight cooling.

The oceans, also, have a pronounced effect on the heat budget of Earth. Water has the highest specific heat of any common substance. That means more heat is needed to raise the temperature of water than to raise the temperature of most other materials. Summers and daylight periods are kept cooler by the water's ability to absorb great amounts of solar energy without the water's temperature being raised significantly. During winter and during darkness, the water slowly gives up large amounts of heat without significant cooling of the water. Thus, the oceans act as a huge temperature buffer and, along with the atmosphere, add a moderating effect to temperature extremes.

Another factor in moderating temperature variations is the rate of Earth's rotation on its axis—one rotation in twenty-four hours. Rotation causes places on Earth to be alternately turned toward and away from the Sun, as though it were on a rotisserie. The relatively rapid rotation prevents places on Earth from overheating or overcooling. If Earth rotated significantly more slowly, so one side were exposed to the Sun for a much longer time, the illuminated side would become considerably hotter, while the dark side would cool down considerably more. For example, the planet Mercury rotates on its axis once in 58.6 Earth days, and it revolves around the Sun in 88 days; in other words, it makes two orbits around the Sun in the same time it completes three rotations on its axis. As a result, a given spot on the planet's surface is exposed to sunlight for 88 Earth days, and then is in darkness for 88 days more. The resultant temperature extremes range from about 430 to -170° Celsius (703 to 103 kelvins). Similarly, a given spot on the surface of the Moon is exposed to sunlight for a little over two Earth weeks and then is in darkness for about two weeks more, resulting in temperatures that range from about 120 to minus 170° Celsius (393 to 103 kelvins) between the day and night sides.

One complete revolution, or orbit, of Earth around the Sun defines the time unit of one year. During a single revolution, Earth rotates on its axis 365.25 times; therefore, there are 365 days in most calendar years, with an extra day every fourth year (leap year). The orbit of Earth around the Sun is an ellipse, which lies in a plane called the ecliptic plane. The Sun is located at one of the two foci of the ellipse; thus the distance of Earth from the Sun varies during the year. The point on the orbit where Earth is closest to the Sun is called perihelion; it occurs on or about January 3 each year at an Earth-Sun distance of about 147 million kilometers. The point on the orbit farthest from from the Sun is called aphelion; it occurs on or about July 4 each year at an Earth-Sun distance of about 152 million kilometers.

This variation in Earth's distance from the Sun does alter the amount of solar radiation that is received by Earth, but it is not the cause of the seasons. Perihelion, when Earth is nearest to the Sun and seemingly when Earth would be the warmest, occurs during winter in the Northern Hemisphere, and aphelion occurs during the Northern Hemisphere's summer. Thus the distance variations are out of phase with the seasons in the Northern Hemisphere, but in phase with seasons in the Southern Hemisphere. In both cases, the distance variations modify seasonal temperatures but do not cause the seasons themselves.

The cause of the seasons is the fact that Earth's axis of rotation is tilted 23.5° from the perpendicular to the ecliptic plane, which is the plane of Earth's orbit around the Sun. The orientation in space of Earth's rotational axis changes only very slowly, so that during one year (one orbit around the Sun) it remains nearly constant in position. As Earth revolves around the Sun, the axis in the Northern Hemisphere is alternately tilted toward and away from the Sun. When Earth's North Pole is tilted toward the Sun, the Northern Hemisphere receives more solar radiation than does the Southern Hemisphere, resulting in summer in the Northern hemisphere and winter in the Southern Hemisphere. When Earth's North Pole is tilted away from the Sun, the opposite occurs.

At the point in the orbit when the North Pole is tilted most directly away from the Sun, the Sun is exactly overhead at noon at the Tropic of Capricorn (23.5° south latitude) on Earth, and the entire area south of the Antarctic Circle experiences continuous daylight. This position in the orbit and the moment of time when it occurs both are referred to as the December solstice, which occurs around December 21 each year. For the Northern Hemisphere, it is the "winter solstice," but for the Southern Hemisphere, it is the "summer solstice." Six months later, when the North Pole is tilted most directly toward the Sun, the Sun is exactly overhead at noon at the Tropic of Cancer (23.5° north latitude) on Earth, and the entire area north of the Arctic Circle experiences continuous daylight. This position in the orbit and the moment in time when it occurs

both are known as the June solstice, which occurs about June 21 each year. Approximately halfway in between the two solstices are the two equinoxes, occurring about March 20 or 21 and September 22 or 23 each year. On the two equinoxes, the Sun is directly overhead at noon at the Equator (0° latitude). Both Northern and Southern Hemispheres receive equal solar radiation then. On the two equinoxes, the periods of daylight and darkness are equal all over the Earth (equinox means "equal night"), and the Sun rises due east and sets due west.

Between March and September, when the Sun is overhead as seen from north of the Equator, sunrise is north of east and sunset is north of west for all locations that experience sunrise and sunset on a particular date, both Northern and Southern Hemispheres. (The places that do not experience sunrise and sunset are those areas near the poles that are experiencing either continuous daylight or continuous darkness on that date.) Sunrise is farthest north of east and sunset is farthest north of west on the June solstice, after which they both begin a southward migration. Between September and March, when the Sun is overhead as seen from south of the Equator, sunrise is south of east and sunset is south of west for all places that experience sunrise and sunset on a particular date, both Northern and Southern Hemispheres. (Again, the places that do not experience sunrise and sunset are those areas near the poles that are experiencing either continuous daylight or continuous darkness on that date.) Sunrise is farthest south of east and sunset farthest south of west on the December solstice, after which they both begin a northward migration to repeat the pattern.

Methods of Study

The seasonal variations in the Sun's apparent daily motion across the sky were noted by many ancient cultures. Various stone structures built hundreds to thousands of years ago around the world—from Stonehenge on England's Salisbury Plain to Caracol in Mexico's Yucatán peninsula to the Bighorn Medicine Wheel high in Wyoming's Bighorn Mountains to Mystery Hill in southern New Hampshire—display alignments pointing toward the rising and setting points of the Sun on the solstices and equinoxes.

More recently it has been determined that the tilt and orientation of Earth's rotational axis and the eccentricity of Earth's elliptical orbit change slowly and cyclically with time. The tilt of Earth's rotational axis relative to a perpendicular to the ecliptic plane (the plane of Earth's orbit around the Sun) now is about 23.5°, but it varies between approximately 21.5 and 24.5° over a cycle of 41,000 years. A greater tilt results in more extreme summer and winter climates, while a smaller tilt means summers are not as hot and winters are not as cold.

Earth's rotational axis also slowly wobbles like that of a giant top, tracing out in space a double cone over a period of 26,000 years. This wobble is due to the gravitational pull of the Moon and the Sun on Earth's equatorial bulge. At the present time, we are closest to the Sun during northern winter and farthest away during northern summer, but as a result of precession, in 13,000 years we will be farthest during northern winters and closest during northern summers, making seasons more severe in the Northern Hemisphere but less severe in the Southern Hemisphere.

Finally, the shape of Earth's orbit around the Sun slowly alternates between being more nearly circular and slightly more elliptical over a period of about 100,000 years due to gravitational perturbations by the other planets. The interplay of all these changes alter slightly the solar radiation received in the Northern and Souther Hemispheres and hence their seasonal climate variations. The Serbian astrophysicist Milutin Milankovi2 was the first to study the effects of these changes and link them to the multiple advances and retreats of large-scale continental glaciation in the Northern Hemisphere during the Pleistocene epoch (the last two million years of geologic time).

Many solar phenomena such as sunspots, prominences, and flares vary in number and frequency of occurrence over a period of about eleven years (the solar activity cycle). This is caused by changes in the Sun's magnetic field, which reverses direction with each solar cycle. Long-term studies of the Sun show that the Sun's activity level varies over timescales of hundreds of years, becoming more or less active. The changes in solar activity seem to be related to changes in Earth's climate, as recorded in old documents and preserved in the width of annual tree rings. Very little solar activity was observed during the 1600's and 1700's, a period known as the Maunder minimum. During this time, Europe and northeastern North America were colder (the so-called Little Ice Age) and western North America experienced prolonged droughts. In contrast, solar activity seems to have been unusually high from sometime in the 1000's until about 1250, a time known as the Medieval Optimum (also known as the medieval grand maximum) when the climate was warmer than it is today. This time marked the height of the Vikings' expansion, when they established colonies in Greenland and Newfoundland. The colonies were abandoned or died out when solar activity declined and the climate turned colder.

Over still longer timescales, theories of stellar evolution applied to the Sun indicate that it has slowly increased in brightness since it formed about 4.5 billion years ago and that it will continue to do so for several billion years more, until it begins to run out of hydrogen fuel in its interior. When that happens, the Sun will expand relatively rapidly to become a red giant star several hundred to more than a thousand times brighter than it is now. These changes will greatly increase Earth's temperature, eventually making it uninhabitable.

Accurate measurements of the length of the day show small, erratic changes in Earth's rotation period, but on average an Earth day is lengthening by about 0.001 second every century. This slowing of Earth's rotation is due to tidal friction. The gravitational effects of the Moon and to a smaller extent the Sun produce tides on Earth, which gradually retard Earth's rotation. To conserve angular momentum, the Moon's distance from Earth is slowly increasing. This has been confirmed by accurately measuring the out-and-back travel time of laser beams bounced off retroreflectors left on the Moon's surface by the Apollo Moon landing missions. These processes will continue until, in the distant future (at least billions of years), Earth's rotation will become tidally locked with the Moon's revolution around Earth, both taking about 47 of our present days. However, this is happening so slowly that the Sun probably will become a red giant first.

Evidence of changes in Earth's heat budget in the past (from the recent past to the ancient past) have come from many disciplines, including history, geology, paleontology, climatology, and astronomy. Satellite studies of Earth over the past several decades have opened many new ways to monitor present conditions and look for predictors of possible future changes. For example, satellite images of the oceans at various wavelengths of the electromagnetic spectrum are analyzed to detect any slight temperature changes over time that may portend changes in Earth's climate. Sensitive satellite-borne instruments measure the intensity of sunlight in remote areas. Satellite imagery also provides an accurate record-base for changes in areas of snow cover in polar regions.

Context

Life on Earth is profoundly dependent upon the relationships between Earth and the Sun. The temperature of Earth is set by a balance between Earth's absorption of electromagnetic energy from the Sun and the subsequent reradiation of that energy from Earth as heat back into space. Life on the planet is dependent on this balance and the moderate temperatures that result.

Rotation influences Earth like a rotisserie, turning the planet so as to expose all sides to the Sun during the twenty-four-hour day for a more even heat. The atmosphere protects Earth from overheating by day and from overcooling at night. Earth's "greenhouse effect" is a result of the atmosphere's ability to trap solar radiation as heat during the day and retard its escape back into space at night, when the Sun is not above the horizon. Earth's heat budget is a product of many factors, not all of which are fully understood. Intense research continues on possible causes and effects of changes in Earth's heat budget. Being able to predict future changes is of prime importance so we can either prepare for them or try to avert them.

John H. Corbet

Further Reading

Ahrens, C. Donald. *Meteorology Today: An Introduction to Weather, Climate, and the Environment.* 8th ed. Florence, Ky.: Brooks/Cole, 2006. This introductory college-level text on meteorology presents a thorough treatment of weather phenomena and explains the seasons and the effects of solar energy on the atmosphere. Written for students with little background in science or mathematics. Includes many illustrations.

Chaisson, Eric, and Steve McMillan. *Astronomy* Today. 6th ed. New York: Addison-Wesley, 2008. Very well-written college-level textbook for introductory astronomy courses. Part of one chapter deals with Earth motions and the seasons; part of another, with solar activity.

Fraknoi, Andrew, David Morrison, and Sidney Wolff. *Voyages to the Stars and Galaxies.* Belmont, Calif.: Brooks/Cole-Thomson Learning, 2006. A well-written, thorough college textbook for introductory astronomy courses. Has sections dealing with sky motions, the seasons, and solar activity.

Freedman, Roger A., and William J. Kaufmann III. *Universe.* 8th ed. New York: W. H. Freeman, 2008. College-level introductory astronomy textbook, thorough and well written. Includes sections on sky motions, the seasons, and solar activity.

Gabler, Robert E., Robert J. Sager, Sheila M. Brazier, and D. L. Wise. *Essentials of Physical Geography.* 8th ed. Florence, Ky.: Brooks/Cole, 2006. A general introductory-level text on physical geography. Covers rotation, revolution, solar energy, and the elements of weather and climate. Well illustrated. Suitable for the general reader.

Harrison, Lucia Carolyn. *Sun, Earth, Time, and Man.* Chicago: Rand McNally, 1960. This book is considered the classic reference for Earth-Sun relations. Although dated, it is an excellent source of information and offers a remarkably extensive coverage of Earth-Sun relations.

Ruddiman, William F. *Earth's Climate: Past and Future.* 2d ed. New York: W. H. Freeman, 2008. A college textbook, suitable for both introductory and upper-level undergraduate courses. Contains much material on climate change and its causes.

Schneider, Stephen E., and Thomas T. Arny. *Pathways to Astronomy.* 2d ed. New York: McGraw-Hill, 2008. Very thorough college textbook for introductory astronomy courses, divided into many short sections on specific topics. Has several sections on the motions of Earth, equinoxes and solstices, and solar activity.

Strahler, Arthur N., and Alan H. Strahler. *Modern Physical Geography.* 4th ed. New York: John Wiley & Sons, 1992. In this general college-level text on physical geography, Earth-Sun relations are discussed within the context of the study of weather and climate. Diagrams are well employed to explain Earth's orbit, rotation, revolution, and axis tilt. Easy to read.

Tarbuck, Edward J., and Frederick K. Lutgens. *Earth: An Introduction to Physical Geology.* Illustrated by Dennis Tasa. 9th ed. Upper Saddle River, N.J.: Pearson Prentice Hall, 2008. Several chapters in this introductory text deal with the solar system. The nature of solar activity and the Earth's motions are explained. Well illustrated and accessible.

EARTH SYSTEM SCIENCE

Categories: Earth; Scientific Methods

Earth system science views the planet Earth as a dynamic, unified system of simultaneous, interacting forces. In particular, Earth system science focuses on achieving a better understanding of the effects of human interaction with the environment.

Overview

A new approach to studying the Earth's systems views them as a set of interacting forces all operating simultaneously rather than separate Earth science disciplines to be studied in isolation. This new and promising viewpoint came about as a result of a growing recognition of the interactive nature of Earth's forces, as exerting influences on one another, as opposed to the idea that these forces act independently. The Earth is a constantly changing world with dramatic tectonic activity, volcanism, mountain building, earthquakes, dynamic oceans, severe storms, and varying climatic patterns and atmospheric conditions. Scientists using this "systems approach" view the Earth as a unified whole, and instead of concentrating attention on one component at a time, they use total global observation methods (attempting to model Earth as a whole) together with numerical modeling.

The Earth systems science approach was first detailed by an Earth System Science Committee (ESSC) appointed by the Advisory Council of the National Aeronautics and Space Administration (NASA). In 1986, the committee completed a three-year study of research opportunities in Earth science and recommended that an integrated, global Earth observation and information system be adopted and in full operation by the mid-1990's. The committee's *Overview Report* was released on June 26 of that year. Requests for the findings of the committee from the National Oceanic and Atmospheric Administration (NOAA) and the National Science Foundation (NSF), along with other federal agencies, have drawn the agencies—especially NASA, NOAA, and NSF—into a scientific alliance. The committee's report outlined immediate needs in several wide-reaching areas: scientific understanding of the entire Earth as a system of interacting components; the ability to predict both natural and human-induced changes in the Earth system; strong, coordinated research and observational programs in NASA, NOAA, and NSF as the core of a major U.S. effort; long-term measurements, from space and from Earth's surface, to describe changes as they occur and as a basis for numerical modeling; modeling, research, and analysis programs to explain the functioning of individual Earth system processes and their interactions; a sequence of specialized space research missions focusing on Earth systems, including the Upper Atmosphere Research Satellite (UARS), the joint

United States/France Ocean Topography Experiment (TOPEX/POSEIDON), the Geopotential Research Mission (GRM), and an Earth-observing system using polar-orbiting platforms planned as part of the U.S. Space Station complex combining NOAA and NASA instrumentation.

Earth system science utilizes new technologies in global observations, space science applications, computer innovations, and quantitative modeling. These new tools of advanced technology allow scientists to probe and learn about the interactions responsible for Earth evolution and global change. Examples of research made possible by new tools are the opportunity to include the effects of global atmospheric motions in models of ocean circulation; the study of volcanic activity as a link between convection in the Earth's mantle and worldwide atmospheric properties; and the tracing of the global carbon cycle through the many transformations of carbon by biological organisms, atmospheric chemical reactions, and the weathering of Earth's solid surface and soils. In addition, recent advances in these technologies

The Orbiting Carbon Observatory which was lost during launch on February 24, 2009, was designed to be the first spacecraft to study Earth's atmospheric carbon dioxide, a main cause of global warming. (NASA/JPL)

have had the immediate practical effect of improving the quality of human life in areas such as weather prediction, agriculture, forestry, navigation, and ocean-resource management.

The goal of Earth system science is to obtain a scientific understanding of the entire Earth system on a global scale by describing how its component parts and their interactions have evolved, how they function, and how they may be expected to continue to evolve on all timescales. This evolution is influenced by human activities—for example, the depletion of the Earth's energy and mineral resources and the alteration of atmospheric chemical composition—that sometimes are easily identified. The overall long-range consequences of these human actions are difficult to predict; the changes do not occur quickly enough for immediate recognition and, indeed, often take decades to evolve fully. The challenge to Earth system science is to develop the capability to predict those changes that will occur in the twenty-first century, both naturally and in response to human activity. To meet this challenge, vigorous investigations are being undertaken that include global observations, information systems built to process global data, and existing numerical models that already are contributing to a detailed understanding of individual Earth components and interactions. Such programs require interdisciplinary research support and interagency cooperation.

Observations from space, the best vantage point from which to obtain the comprehensive global data required to discriminate among worldwide processes operating on both long and short timescales, are essential to the study of the Earth as a system. Rapid variations in atmospheric and ocean properties, and the global effects of volcanic eruptions, ocean circulations, and motions of the Earth's crustal plates are examples of such processes. The Space Science Board of the National Academy of Sciences recommended orbital observation as a major method of global study; the Earth System Science Committee accepted the recommendations and expanded on them. Of particular value are NASA and NOAA satellites already on station in orbit, such as the Laser Geodynamics Satellites, which employ laser ranging to measure motions and deformations of Earth's crustal plates. Weather satellites already have supplied a sizable fund of data about the atmosphere and oceans, facilitating a good start on numerical modeling of weather variations. Other programs that have yielded coordinated studies of specific Earth system processes include the Earth Radiation Budget Experiment (1984), the Laser Geodynamics Satellites (1976 and 1983), the Navy Remote Ocean Sensing System (1985), and the Upper Atmosphere Research Satellite (1982).

In order to implement the full measure of the Earth system science concept, advanced information systems are needed to process global data and allow analysis, interpretation, and quantitative modeling. Also required is the implementation of additional satellite observations that yield ocean color imaging, scanning radar altimeters for surface topography, and atmospheric monitors. In addition, vigorous programs of ground-level measurements are needed to complement, validate, and interpret the global observations from space. International cooperation is essential to the success of Earth system science; the development of management policies and mechanisms are required to encourage cooperation among agencies around the globe in order to ensure the coordination necessary for a truly worldwide study of the Earth. A number of major international research programs are already operating, such as the World Climate Research Program, sponsored by the International Council of Scientific Unions and the World Meteorological Organization. To accomplish the many objectives of these programs, the Earth System Science Committee recommends two specific goals in which the three major U.S. agencies—NASA, NOAA, and NSF—must work closely together. The first goal is to establish and develop the advanced information systems and management structures required by Earth system science as a cooperative venture, and the second is to build close cooperation in programs of basic research.

Methods of Study

The most significant tools for global observation are Earth-orbiting satellites that can precisely measure large areas of the Earth at one time. Meteorological satellites, for example, gather enormous amounts of data about temperature, weather patterns and forces, and atmospheric changes and components; instruments aboard these satellites can gather data on and monitor variations of climate and storm systems, adding to the growing fund of global information. These satellites are placed in geosynchronous orbit at an altitude of 35,000 kilometers over the equator; at that altitude their orbital period is the same as the Earth's rotation—one day—so they remain over the same spot on Earth and can continuously monitor the same region.

Earth observation satellites, working in the infrared band of the spectrum, allow scientists to gather imagery and information about volcanic activity, earthquakes, geological formations, mineral resources, and geographic changes to provide still another perspective on the Earth. Orbiting the Earth from pole to pole many times a day, they are able to make a record of large sections of the Earth in a twenty-four-hour period as the Earth rotates underneath the satellite's orbital path. Earth observation satellites also carry instruments that measure temperatures, record cloud cover, and monitor catastrophic changes.

Other satellites measure ocean dynamics such as the temperature of large sections of seas and oceans, wave action, ocean water content, and relationships between water and the land it touches. Special instruments aboard these satellites are designed to monitor ice conditions and snowfall at sea and watch for changes in polar regions.

Still other spacecraft carry radar-imaging devices to measure precise distances and relationships between land features. The International Space Station has, as one of its most important objectives, the function of a permanently orbiting platform on which both humans and unattended instruments can work over long periods of time to monitor Earth activities and topography. The space station will be able to contribute large amounts of data because it can function both as information gatherer and processor using advanced onboard automated equipment such as specialized computers.

Although much of the instrumentation for Earth system science will be space-borne, much of it also will have to be ground-based, at the sites where data need to be gathered: near volcanoes, earthquakes, hurricanes, tornadoes, and thunderstorms, for example. Such phenomena must be measured on the ground to determine their effects on other Earth-surface processes. Ground data can then be compared and synthesized with data gathered from space to offer a broader view.

One of the most valuable of tools is the computer, for the receipt, storage, retrieval, analysis, and supply of large quantities of information. Ground-based and space-borne computers work in conjunction with each other for the comparison and large-scale analysis of data, which can be networked to any place on Earth. Computers also are used to generate theoretical models of various kinds of processes. By feeding weather data from the past hundred years into a computer, for example, scientists can begin to construct long-term models of weather patterns and global changes in climate and precipitation. Another study method is the creation and management of global information systems into which is fed data from countries all over the world; all nations can retrieve data for their own research as well as input data to add to the ongoing process of worldwide data analysis.

Context

The new methodology of Earth system science offers the opportunity to study the Earth from a more integrated perspective and to raise public awareness of the human practices that are affecting the planet. It is important that citizens of the twenty-first century understand the forces and processes that can cause global changes because, individually and collectively, they are contributors to those changes. Human contributions include continued clear-cutting of vast forest areas, thus inviting massive deforestation (destruction of forests); removal of protective trees and underbrush from areas adjacent to desert areas, thus encouraging rampant desertification (the spread of desert conditions); and pollution of the atmosphere and waterways. Over time, these practices can slowly deplete Earth's natural resources and upset the fragile balance of nature worldwide. Human activities can trigger events that could cause long-term environmental damage.

Thomas W. Becker

Further Reading

Asrar, Ghassem. *EOS: Science Strategy for the Earth Observing System*. New York: American Institute of Physics, 1994. Describes the Earth Observing System program: its investigations, capabilities, and educational activities. For undergraduates and science readers.

Earth System Science Committee. *An Integrated Global Earth Observation and Information System to Be in Full Operation by the Mid-1990's*. Boulder, Colo.: University Corporation for Atmospheric Research, 1986. Written by the key people who created the method, a basic, concise presentation of the Earth Observation and Information System.

Kump, Lee R., James Kasting, and Robert Crane. *The Earth System*. Upper Saddle River, N.J.: Prentice Hall, 2003. An introductory work for those new to the Earth system field. Addresses the carbon cycle and events in Earth's history that help explain current global changes. For a general audience.

MacKenzie, Fred T. *Our Changing Planet: An Introduction to Earth System Science and Global Environmental Change*. Upper Saddle River, N.J.: Prentice Hall, 2002. Aimed at nonscientists, this volume covers all areas of Earth system science, including global change associated with both natural and human sources.

Matthews, Samuel W. "This Changing Earth." *National Geographic* 143 (January, 1973): 1-37. One of the earliest articles to describe the Earth's dynamic processes in a language that the public could readily understand. Addresses plate tectonics and takes the reader on a historic tour of the development of modern Earth science. Superb diagrams and supportive photography.

National Aeronautics and Space Administration Advisory Council. *Earth System Science Overview*. Washington, D.C.: Government Printing Office, 1986. This fifty-page document details in easy-to-understand language all the intricate natural mechanisms at work on the planet. Describes the entire Earth system science concept and outlines how the discipline's tools and methods will be brought together to focus on a global data-gathering, archiving of information, and international cooperative efforts. For high school students and general readers.

Skinner, Brian J. *The Blue Planet: An Introduction to Earth System Science*. New York: John Wiley, 1995. Good introduction to the field of Earth system science. Contains a series of essays covering both methods of research and current progress, some written by leading scientists. For a general audience.

GEMINI PROGRAM

Category: Space Exploration and Flight

The Gemini Program placed humans into Earth orbit and taught astronauts how to track, maneuver, and control orbiting spacecraft; dock with other orbiting vehicles; and reenter Earth's atmosphere and land at specified locations, all necessary to the execution of an Apollo mission.

Overview

Prior to the formation of the National Aeronautics and Space Administration (NASA), a number of crewed space concepts had been investigated within the military. After the Soviet Union launched the world's first artificial satellite, Sputnik 1, on October 4, 1957, the U.S. reaction could easily be described as one of panic, with fear centering on the suspicion that the Soviet Union would assume technological leadership over the free world.

The United States' first attempt to send a satellite into space failed miserably. Vanguard 1 blew up on the launch pad in December, 1957, before the eyes of the world. On January 31, 1958, an Army-based group including

Wernher von Braun successfully placed Explorer 1 into orbit. Within months, President Dwight D. Eisenhower, with Congressional approval, created NASA as a civilian space agency. Its first major endeavor was the Mercury project, the goal of which was to send an astronaut into orbit before the Russians. However, the Soviets scored another major first when, on April 12, 1961, they launched cosmonaut Yuri Gagarin into space. He completed one Earth orbit before safely returning to Earth, landing within the Soviet Union. The Soviets led the space race thanks to their proficiency with heavy-lift boosters, a strength they continued to exploit for many years. This strength played a major role in Gagarin's achieving orbit before NASA's Mercury astronauts.

On May 5, 1961, Alan Shepard became the first American to enter space. Shepard launched atop a Redstone rocket, which did not have sufficient thrust to lift his Mercury capsule into orbit. Shepard flew a fifteen-minute-long suborbital profile, arcing up to 115 miles altitude and splashing down in the Atlantic Ocean off the coast of Cape Canaveral. Just three weeks later, President John F. Kennedy committed NASA to sending a man to the Moon and back before the year 1970.

To achieve this goal, NASA initiated the Apollo Program. However, the proposed three-man Apollo vehicle was far too big a step over the existing primitive single-astronaut Mercury spacecraft that could orbit the earth for only a brief period. Further, the Mercury spacecraft was incapable of orbital maneuvers of the type necessary for achieving Apollo's goal using what was termed a lunar orbit rendezvous (LOR) technique. LOR involved having the Apollo spacecraft separate into two portions, one that remained in lunar orbit and another that took two astronauts wearing protective pressure suits down to the lunar surface. This separation necessitated a rendezvous after lunar exploration, so that all three astronauts could reunite for the journey back to Earth. Apollo missions would last between eight and fourteen days and would require very precise reentry maneuvers in order to bring the crew safely through a narrow corridor in the earth's atmosphere where the spacecraft would survive reentry heating.

The Gemini Program was therefore developed as an interim means whereby all of the techniques necessary for Apollo missions could be assessed and refined in low-Earth orbit. Gemini astronauts would build up experience with orbital maneuvering, living for prolonged periods of time in weightlessness, rendezvous and docking separately launched target vehicles, and controlled reentries involving splashing down near recovery forces. Whereas the Mercury spacecraft carried one astronaut and the Apollo spacecraft would carry three, Gemini was designed to build on the Mercury experience with a modular spacecraft capable of flying two astronauts for up to two weeks in duration.

Gemini Flight Operations

The first two Gemini missions were uncrewed test flights to rate the Titan II launch vehicle and to qualify critical Gemini spacecraft systems. Gemini 1 was launched successfully from Cape Kennedy on April 8, 1964. A Gemini spacecraft mock-up was placed in orbit as a result of proper Titan II booster performance. The spacecraft was meant neither to be separated from the booster's second stage nor returned to Earth.

Gemini 2, which included the first fully operational test spacecraft, was launched on January 19, 1965. It followed a suborbital profile designed to stress the Gemini spacecraft heat shield's ability to manage reentry heating. The spacecraft was safely recovered from the Atlantic Ocean.

Gemini 3 launched on March 23, 1965, with astronauts Virgil "Gus" Grissom and John W. Young aboard. Over the course of three orbits, the astronauts performed three different maneuvers to change the spacecraft's orbit, the first time that orbital maneuvers were executed on a crewed spacecraft. Russian cosmonauts had flown aboard Vostok and Voshkod capsules that had been largely automated in nature.

Just prior to the Gemini 3 mission, the Russians had scored a major advance when cosmonaut Alexei Leonov departed his Voshkod 2 spacecraft for a brief walk in space. The next Gemini mission, Gemini IV, attempted the first extravehicular activity (EVA) performed from a NASA spacecraft. Gemini IV launched on June 3, 1965, with astronauts James A. McDivitt and Edward White aboard. McDivitt attempted to fly close to the spent Titan II booster's second stage after spacecraft

separation, consuming a great deal of fuel in the process. Because the mission was meant to last four days, a NASA first, it was decided to halt that maneuver to save fuel. On the mission's third orbit, White opened his hatch and proceeded to exit from the spacecraft using a small gas-powered thruster gun to move about while remaining tethered to Gemini IV by a life-support umbilical. White quickly depleted his gas supply, but he spent a total of twenty-three minutes floating about before returning to Gemini IV's cabin. After four days, McDivitt flew a manually controlled rolling reentry, and Gemini IV splashed down within range of recovery forces in the Atlantic Ocean.

Astronaut Gordon Cooper is recovered after the splashdown of the Gemini V space capsule on August 29, 1965. (NASA)

Gemini IV began NASA's evolutionary buildup toward a two-week-long mission, and its astronauts spent as much total time in space as had that astronauts of all previous NASA crewed flights combined.

Gemini V launched on August 21, 1965, with astronauts L. Gordon Cooper and Charles "Pete" Conrad on board. This mission marked the first use of fuel cells utilizing liquid oxygen and hydrogen to produce electrical power. Problems with systems associated with the fuel cells surfaced early in flight, forcing the cancellation of a rendezvous exercise and a powering-down of the spacecraft. Those pressure problems diminished later in the flight, and Cooper and Conrad were able to remain aloft in Gemini V for nearly eight full days before returning to Earth, splashing down in the Atlantic Ocean with no major physiological problems encountered during their record-setting flight.

The next Gemini mission was a planned rendezvous and docking with an uncrewed Atlas-Agena docking target. The original flight plan called for the liftoff of the Agena docking target on top of an Atlas rocket about an hour before that of the Gemini VI spacecraft. Once the Agena had reached the correct orbit, Gemini VI would be launched. Over a period of time, Gemini VI would catch, rendezvous, and join or dock with the Agena. Both of these vehicles would be launched from Cape Kennedy, Florida, from separate launch pads.

The first launch attempt of Gemini VI was made on October 22, 1965. Unfortunately, the Agena suffered a failure shortly after it separated from the Atlas booster and was lost. Astronauts Walter Schirra and Thomas Stafford were already in their spacecraft, but, with no Agena target in space, Gemini VI was scrubbed in favor of proceeding with Gemini VII, a two-week-long flight of astronauts Frank Borman and Jim Lovell in December, 1965. Gemini Program managers decided to alter Gemini VI's mission, renaming it Gemini VI-A, and to use Gemini VII as a target with which to rendezvous if Cape Kennedy personnel could refurbish the launchpad sufficiently quickly following Gemini VII's launch to permit a second Gemini liftoff within a two-week period.

Gemini VII was launched on December 4, 1965, and entered stable orbit. An attempt to launch Gemini VI-A on December 12 resulted in an engine shutdown on the pad, but the crew was safe. Then, on December 15 with the booster refurbished, Gemini VI-A lifted off and began a four-orbit chase, closing to within a foot of Gemini VII's nose. Even without docking, this dual flight verified the capability of astronauts to execute the maneuvers needed

for Apollo. Also, the Gemini VII crew proved that astronauts could survive weightlessness during the longest Apollo flights. Gemini VI-A returned to Earth, splashing down in the Atlantic Ocean on December 16. Gemini VII executed its reentry on December 18, landing close to the same recovery vessel that had recovered Gemini VI-A.

Gemini VIII launched on March 16, 1966, with astronauts Neil Armstrong and David Scott on board. Their liftoff came one orbit after an Atlas booster delivered an Agena to orbit. Armstrong and Scott executed a rendezvous over the course of four orbits and docked to their Agena vehicle. Within one half-hour, Gemini VIII and its Agena entered a rolling motion that threatened structural stability. Armstrong undocked and backed away from the Agena. As the problem involved a Gemini VIII thruster firing uncontrollably, the roll rate increased, forcing Armstrong to regain control by firing other thrusters dedicated for reentry. Gemini VIII had to be terminated early, and Armstrong and Scott splashed down in a backup recovery zone in the Pacific Ocean after only seven orbits.

After Gemini IX's Agena target, launched on May 17, 1966, failed to reach orbit, the mission was postponed. An alternate target called an Augmented Target Docking Adapter (ATDA) was launched on June 1, 1966, but the crewed Gemini flight, now renamed IX-A, could not follow. Two days later, astronauts Stafford and Eugene Cernan launched from Pad 19. When Gemini IX-A approached the ATDA after a three-orbit rendezvous, the astronauts found the ATDA's forward shroud had not cleanly separated from the docking mechanism. They used the ATDA for several different rendezvous exercises, but no docking was possible. Cernan attempted a spacewalk meant to last one full orbit, but he ran into difficulties working on a jet backpack he intended to test-fly up to 100 feet away from Gemini IX-A. His visor fogged over, and he had to terminate the EVA and return to the cabin. Gemini IX-A splashed down in the Atlantic Ocean after 45 orbits.

Gemini X included a pair of rendezvous exercises involving the Agena VIII target vehicle and an Agena launched one orbit before astronauts John W. Young and Michael Collins. Gemini X and Agena X were launched on July 18, 1966. After docking Gemini X and Agena X together, a rocket firing of Agena X's main engine propelled the docked complex toward rendezvous with Agena VIII. Collins performed a tethered EVA in the proximity of Agena VIII and also performed another spacewalk while standing up in his seat to perform astronomical observations. This marked the first time that all major objectives of the Gemini Program were demonstrated in one single mission. Gemini X splashed down in the Atlantic Ocean on July 21.

Gemini XI and its Agena were launched on September 9, 1966. Astronauts Conrad and Richard F. Gordon completed a rendezvous on their first orbit and docked to Agena XI. Gordon performed a stand-up spacewalk and an umbilical EVA, the latter requiring early termination, after Gordon overstressed his life-support chest pack's ability to keep him cool. Using Agena XI's propulsion system, Conrad and Gordon were able to temporarily boost their spacecraft up to a record 850-mile altitude. They splashed down in the Atlantic Ocean after executing the first computer-controlled reentry.

Gemini XII and its Agena were launched on November 11, 1966. Astronauts Lovell and Edwin "Buzz" Aldrin performed several rendezvous and docking exercises, expanding NASA's experience base. Perhaps the most important aspect of the final Gemini mission involved Aldrin's three periods of spacewalking, two of the stand-up variety and one umbilical. He spent a total of almost five and one-half hours outside the spacecraft and demonstrated methods that overcame problems encountered by earlier Gemini spacewalkers. Gemini XII landed on November 15 in the Atlantic Ocean.

Historical Context

From March, 1965, to November, 1966, Gemini astronauts flew ten crewed missions, greatly expanding NASA's crewed space flight experience beyond that of the original seven Mercury astronauts. During that period, not a single Russian cosmonaut flew in space, and NASA finally overcame the early Soviet lead in space technology. Gemini flights investigated virtually all aspects of an Apollo mission and laid the foundation for the successful achievement of a crewed lunar landing in July, 1969.

David G. Fisher

Further Reading

Collins, Michael. *Liftoff.* New York: Grove Press, 1988. Provides an astronaut's perspective of the Gemini and Apollo Programs.

Hacker, Barton C., and James M. Grimwood. *On the Shoulders of Titans: A History of Project Gemini.* Washington, D.C.: Government Printing Office, 1977. Provides a thorough historical chronicle of program engineering and management evolution.

Schirra, Walter M., Jr., with Richard N. Billings. *Schirra's Space.* Boston: Quinlan Press, 1988. Provides an astronaut's perspective of NASA's crewed space flight programs from Mercury through Apollo.

GLOBAL ENERGY BALANCE

Category: Astronomy

The global energy balance is the overall inflow and outflow of energy to and from Earth. The different processes that control this balance are atmospheric reflectance, absorption by the atmosphere and Earth's surface, and reradiation of energy to space by Earth itself. Electromagnetic energy arrives in the form of solar radiation that passes through the atmosphere and interacts with the surface.

Overview

When electromagnetic energy hits an object, such as Earth, three processes can occur. Transmission occurs when the energy passes through an object without affecting it. Reflection occurs when the energy bounces off the object, and absorption occurs when the energy is taken up by the object. Incoming solar radiation exists at all wavelengths, but the dominant intensity is within the visible band (400-700 nanometers), at 550 nanometers. Of this energy, 30 percent is reflected from the atmosphere back into space, 25 percent is absorbed by the atmosphere, and the remaining 45 percent is absorbed by Earth's surface.

The energy absorbed by Earth's surface is converted to thermal energy and raises the surface temperature. Because the Earth itself is a warm object, it also emits electromagnetic energy to space. The energy emitted by Earth occurs in longer wavelengths than incoming solar radiation, with the maximum intensity at 10 millimeters. Of this energy, 70 percent passes through the atmosphere and is transmitted to space. However, 30 percent is reabsorbed by atmospheric greenhouse gases (GHGs), including carbon dioxide (CO_2), methane, nitrous oxide, and water vapor. This reabsorbed energy contributes to further warming of Earth. The balance between the amount of incoming solar radiation and the amount of reflected, absorbed, and transmitted energy controls global wind patterns and oceanic circulation. Shifts in this balance determine the overall warming and cooling of the planet.

Significance for Climate Change

Changes to the global energy balance, which is affected by both natural and anthropogenic processes, are a fundamental cause of climate change. Processes that affect the global energy balance are often intertwined with one another, resulting in various feedback mechanisms that can enhance or reduce the effects of the original change. Natural changes in the global energy balance can be caused by variations in Earth's orbit or in the output of solar radiation—both of which affect the amount of energy coming into Earth.

The shape of Earth's orbit and the tilt of Earth's axis change slightly over timescales ranging from 11,000 to 100,000 years. These changes can cause the Earth to be somewhat closer to or further away from the Sun, leading to a relative increase or decrease in the amount of solar radiation that reaches Earth. In the geologic past, these types of changes are manifested as climate shifts on timescales of millions of years, known as greenhouse and icehouse conditions. During greenhouse times, Earth is ice-free because of high concentrations of GHGs. During icehouse times, Earth is covered by variable proportions of ice, and the climate shifts between interglacial times (less ice) and glacial times (larger continental ice sheets) on timescales of tens of thousands of years. The most recent ice age peaked at the Last Glacial Maximum, approximately 20,000 years ago. Approximately 850-630 million years ago, Earth may have been completely covered by ice, creating a so-called Snowball Earth. This hypothesis, however, remains highly controversial among scientists. Blocking some proportion of incoming solar radiation remains a goal of certain geoengineering proposals as a means to prevent future global warming.

Changes in GHG concentrations also affect the global energy balance—and therefore, global temperature—because of changes in the amount of reradiated energy trapped on Earth. Volcanoes are natural sources of GHGs, and geologic evidence suggests that there is a correlation between widespread volcanism and greenhouse times in the past. Anthropogenic changes in the global energy balance are thought to be caused by the addition of GHGs to the atmosphere. The most notable of these is the production of CO_2 through the burning of fossil fuels, although the production of methane should not be discounted. Increased concentrations of GHGs in the atmosphere lead to increased absorption of energy radiated from Earth, which, in turn, leads to further warming of the surface. The Intergovernmental Panel on Climate Change (IPCC) estimates that anthropogenic changes in atmospheric CO_2 concentrations have resulted in a global temperature increase of 0.74° Celsius between 1905 and 2005. Removing CO_2 from the atmosphere and sequestering it in reservoirs such as the deep ocean, ocean sediments, or rock formations constitutes a second geoengineering approach to reduce future global warming.

Anna M. Cruse

Further Reading

Intergovernmental Panel on Climate Change. *Climate Change, 2007—The Physical Science Basis: Contribution of Working Group I to the Fourth Assessment Report of the Intergovernmental Panel on Climate Change*. Edited by Susan Solomon et al. New York: Cambridge University Press, 2007. Comprehensive treatment of the causes of climate change, written for a wide audience. Figures, illustrations, glossary, index, references.

Kunzig, Robert. "A Sunshade for Planet Earth." *Scientific American* 299, no. 5 (November, 2008): 46-55. Explains the basic premises of geoengineering and discusses the potential pros and cons of a variety of geoengineering approaches. Figures.

Rapp, Donald. *Assessing Climate Change: Temperatures, Solar Radiation, and Heat Balance*. New York: Springer, 2008. Presents evidence for climate change—both natural and anthropogenic. Rapp is a systems engineer who attempts to consider the Earth system as a whole. Figures, tables, index, references.

Walker, Gabrielle. *Snowball Earth: The Story of the Great Global Catastrophe That Spawned Life as We Know It*. New York: Crown, 2003. A presentation of the Snowball Earth hypothesis, as developed by Paul Hoffman, for nongeologists. Combines geological theory and data with stories of Hoffman's travels to remote locations to conduct his research. Index, references.

HABITABLE ZONES

Category: Life in the Solar System

Habitable zones are the places beyond Earth where there is the best chance of finding life. As such, they are a major focus of scientific consideration and investigation that inspire exploratory efforts.

Overview

At root, the idea of a habitable zone around a star is simple. A planet must not be too close to a star to be too hot for life. On the other hand, it must not be too far from the star to be too cold for life. The spherical shell around a star where a planet will be "just right" for life is the habitable zone of that star. Another name is the circumstellar habitable zone; this name distinguishes it from the galactic habitable zone, the region of a galaxy most favorable to life.

Unfortunately, it is not a simple matter to calculate the limits of this shell for a given star. However, using Earth and the Sun as the basic measure and noting that energy concentration of light from a star decreases as the square of the distance, scientists can predict the rough average radius of the habitable zone around a star. The result is simply the square root of the ratio of the stellar brightness to that of the Sun. The radius will then be expressed in astronomic units (AU), where 1 AU is the average distance of Earth from the Sun. The brightness of the star must be rated at a standard distance, known as its bolometric luminosity.

To complicate matters, a planet might not stay in the habitable zone. To be habitable, a planet should have an orbit that remains in the habitable zone for billions of years. Only planets with relatively circular orbits can do that. Also, the habitable zone moves away from a star as that star ages and grows hotter. Stars are not all the same. Some burn their fuel quickly and do not last for billions of years. The habitable zone of a bright, hot star will not be the same as that of a dim, cool one. Thus, the habitable zone is determined by the star, the planet, and the life-form under consideration.

The type of life involved is, of course, a critical consideration. We cannot discuss habitable zones without first establishing the expectations we have of life around a star. We can identify three very basic requirements for life. First, living things have bodies. Second, a life-form uses a flow of nutrients and energy to sustain its body and bodily processes. Third, life reproduces itself. Reproduction requires bodies (and, most likely, molecules) able to retain the complex and detailed information required for constructing more living forms. Life may be more than this, but it will surely never be less.

With these criteria we find, by an argument too extended to give here, that life in a habitable zone will be water- and carbon-chemistry-based. The habitable zone, then, can be calculated based on the requirement that a planet in the zone will be able to hold water in liquid form long-term. The actual calculation is complicated by many factors; chief among them is the fact that water vapor is a major greenhouse gas. Hence, one cannot simply find the incident energy from the star at various distances because the presence of water retains the heat supplied by the star and thereby expands the habitable zone. An early estimate by Michael Hart had the habitable zone of our Sun between 99 and 105 percent of the Earth's current distance. This was too conservative; a more likely estimate is 95 to 137 percent.

The habitable zone is unique to the star and is determined by the stellar mass and, to a lesser extent, the age of the star. In terms of spectral classes, stars such as Earth's Sun (in class G) and some K- and F-class stars can have habitable zones. The range also corresponds to the stellar surface temperature range of a bit less than 4,500 kelvins (K) to a bit above 7,000 kelvins. Our Sun, at 5,777 kelvins, is in the middle of this range. Stars with a mass 20 percent or more greater than that of the Sun (that is, 1.2 M_S) will not have habitable zones, because they emit deadly amounts of ultraviolet radiation (UV) along with their visible and infrared radiation (IR). UV destroys water molecules and, at high intensity, will eventually strip a planet of the water critical for life. About 1 percent of stars are so large that they consume their fuel and die long before life can form. Indeed, all stars larger than 1.5 M_S would turn into red giant stars and swallow up any life-bearing planets around them before intelligent life could appear. Stars with more than ten times the mass of our Sun are so intensely bright that planets cannot form around them, because light creates pressure on anything it strikes. This radiation pressure is usually too small to matter, but for these very large stars it is great enough that all the material around the star that might eventually form planets is pushed away from the star and is disbursed too quickly for planets to form.

On the other hand, stars with less than about 0.80 M_S do not produce enough high-energy UV light to support life on any planet; their UV output is insufficient for important atmospheric effects such as ozone creation. Any planet close enough to a star of less than about 0.65 M_S to receive sufficient heat will be so gripped by the stellar gravity that it will show the star one face, as the Moon does our Earth. If this is not the case, an effect called spin-orbit coupling will almost certainly force the rotation rate of the planet to be almost as slow as the planetary year, thus frying the planet on one slowly changing side while freezing it on the other. Mercury is such a case. It revolves around the Sun in 88 days but rotates once every 58.7 days, exactly two-thirds of the orbital period. Both these effects are due to the fact that no planet is perfectly spherical. In either case, the planetary face toward the star will be too hot for water, the side away from the star too cold.

Another factor in habitability is variability of stellar output. Our Sun has an eleven-year sunspot intensity cycle that causes a variation in solar luminosity of about 0.1 percent. However, 18 Scorpii, an almost identical star in the constellation Scorpius with a mass of 1.03 M_s and a temperature of 5,789 K, has a much greater variability over a 9 to perhaps 13 year cycle. If great enough, this would make its habitable zone move in and out rapidly thereby negating its benefits for any planet in a basically fixed orbit.

Knowledge Gained

The idea of a circumstellar habitable zone has stimulated wide-ranging research resulting in a significant extension of our knowledge of planetary systems generally, as well as of our own solar system in particular. A circumstellar habitable zone imposes quite severe limitations on where best to look for life in the universe. Responses to these limitations are likewise limited. One either accepts the limitations, at least tentatively, and looks for suitable planets around only suitable stars, or one must in some way challenge the limitations.

Tentative acceptance of the limitations takes us in the direction of what has become a successful search for exoplanets, planets orbiting stars other than our Sun. The list is large and growing. The primary technique used in this search detects stellar motion due to the stellar reaction to the orbital motion of the planet. Since large planets create more stellar reaction that is more easily detected, this technique is biased toward discovering large planets. It is no surprise then that most of the known exoplanets are large. It is a bit of a surprise that they tend to be relatively close to their stars and, hence, are sometimes called "hot Jupiters." If this trend continues, it may require revisions in the theories of how planetary systems form. Hot Jupiters are not expected to harbor life even if they are in a habitable zone.

Another puzzling result of these searches is that exoplanets seem to prefer highly elliptical orbits compared with those in our solar system. Such orbits are risky in that they may take the planet out of the habitable zone annually. On a more positive note, the work on exoplanets has confirmed that, as expected, planets tend strongly to be found around stars with high metal content. (In this context, "metal" means any element other than hydrogen or helium.)

Challenges to the idea of the circumstellar habitable zone have either been attempts to show there are niches of habitability outside the habitable zone or efforts to extend the habitable zone in size or to more types of stars, especially to red dwarfs. This later direction seems promising in light of the discovery of planets around the red dwarf Gliese 581. One of them, Gliese 581 c, is said to be the smallest planet yet discovered in the habitable zone of another star. That, of course, assumes that a red dwarf has a habitable zone.

Looking for niches of habitability in our solar system—and, hence, potentially elsewhere—offers the possibility of confirmation by direct examination in the not too distant future and is accordingly fairly popular. Thus, attention has become focused on Mars and some large satellites of Jupiter and Saturn.

Mars has received the most attention, as attested by missions such as the Mars Exploration Rover (which began operating on the Martian surface in early 2004). The Cassini-Huygens mission to Saturn (which entered into orbit around Saturn in mid-2004) included flybys of Titan, revealing its liquid methane oceans and dense atmosphere, while the Galileo mission to Jupiter in the late 1990's gathered a great deal of data on two of Jupiter's satellites, volcanic Io and Europa, whose deep ice sheet appears to have water in liquid form beneath it.

Context

The dream of "other races of men" on other worlds has been the currency of cosmological speculation at least since the ancient Greek atomists. Men on the Moon were described by the ancient Pythagoreans and in the seventeenth century by Johannes Kepler, and even the eighteenth century philosopher Immanuel Kant gave opinions on the inhabitants of Mars, Venus, and Jupiter. Modern science has tried to inform and thereby reduce this speculation. The concept of a habitable zone is a product of this effort, although it imposes limitations that would, no doubt, have disappointed earlier enthusiasts.

Enthusiasm for finding life elsewhere in the universe is by no means dead. The high profile of the Search for Extraterrestrial Intelligence (SETI) and the advent of the new academic discipline of astrobiology are proof of that. Both of these developments are inextricably connected with the concept of habitable zones and are all but inconceivable without it. The prospect of habitable zones has also stimulated thinking and research in other areas. One such development is the idea of a galactic habitable zone.

The concept of habitable zones also connects to larger cosmological issues, such as questions of the "fine tuning" of the universe that makes life possible somewhere in the universe and the related issue of the anthropic principle, the notion that the universe must contain conditions that allow for the existence of an observing intelligent life-form.

John A. Cramer

Further Reading

Aczel, Amir D. *Probability 1*. New York: Harcourt Brace & Company, 1998. Aczel argues that the large number of stars outweighs the limitations of habitable zones to the point where intelligent life must occur throughout the universe.

Cohen, Jack, and Ian Stewart. *Evolving the Alien: The Science of Extraterrestrial Life*. London: Ebury Press, 2002. Cohen and Stewart dispute that alien life will be similar enough to terrestrial forms to frame a meaningful idea of a habitable zone. They also argue the case for various niches of habitability.

Cramer, John A. *How Alien Would Aliens Be?* Lincoln, Nebr.: Writers Club Press, 2001. The first half of the book shows how physical constraints limit where intelligent life might be found in the universe. Hence, it surveys many of the limitations that lead to the idea of a habitable zone and the possibility of habitable niches.

Dole, Stephen H. *Habitable Planets for Man*. 2d ed. New York: Elsevier, 1970. Something of a classic on habitable planets, this is one of the earliest discussions of habitable places for human colonization. It gives a good if somewhat dated account of what makes a place habitable for intelligent life, a more restrictive notion than a habitable zone for any life.

Gonzalez, Guillermo, and Jay W. Richards. *The Privileged Planet: How Our Place in the Cosmos Is Designed for Discovery*. Washington, D.C.: Regnery, 2004. Gonzalez and Richards consider the idea that planetary habitability may be connected with the planet's suitability as a platform for observing the universe.

Grinspoon, David. *Lonely Planets: The Natural Philosophy of Alien Life*. New York: HarperCollins, 2004. This is a wide-ranging and readable book covering habitable zones and many related topics.

Ward, Peter, and Donald Brownlee. *Rare Earth: Why Complex Life Is Uncommon in the Universe*. New York: Springer, 2000. Ward and Brownlee make the
case that the limitations on habitable zones are severe to the point of making planets like Earth quite rare.

INTERPLANETARY TRAVEL

Category: Space, Time, and Distance.

Interplanetary travel can be defined as any spaceflight—manned or remotely guided—to the various bodies of the solar system, including planets, their satellites, and asteroids. Such space exploration required new mathematics to plan trajectories and navigate in space, as well as to measure and to analyze massive amounts of data. These flights have had a great societal impact and have radically changed human attitudes toward the outer space surrounding the Earth.

Overview

A scientific possibility of interplanetary travel was discussed for centuries after Isaac Newton wrote *Principia* in 1687, in which he unified terrestrial and celestial dynamics by discovering the force of gravity as an important source of motion, including the movement of celestial bodies. Step by step, an important new mathematical branch of astronomy emerged and received the title "celestial mechanics." In its formative days, celestial mechanics played an outstanding role in the progress of mathematics, demanding and inspiring novel and efficient mathematical tools. Among the pioneers of celestial mechanics were prominent mathematicians such as Leonhard Euler (1707–1783), Alexis-Claude Clairaut (1717–1765), and Joseph-Louis Lagrange (1736–1813). Today, the branch of celestial mechanics dedicated to spaceflight is usually termed astrodynamics.

For many years following Newton's discovery, the topic of interplanetary travels mainly remained the subject of science fiction writers. In the nineteenth century, among the most influential science fiction writers were Jules Verne (1828–1905) with his books *From the Earth to the Moon* and *All Around the Moon* and H. G. Wells (1866–1946) with his book *War of the Worlds*. Verne's work contained a great deal of mathematics discussion, much of which was reasonably accurate based on the knowledge of the time.

To put interplanetary travel into practice, it was necessary to realize some significant preconditions, including designing spacecraft with the capacity for maneuvering, designing technologies for boosters to reach escape velocity, developing a theoretical base for space navigation, and creating systems for long-distance radio communications. These technological developments were not made until the beginning of the space era in 1957.

Mathematical Development

From a mathematical viewpoint, the most interesting part of interplanetary travel is space navigation. An appropriate example of a solution with respect to navigational problems is the Hohmann transfer orbit. In 1925, Walter Hohmann calculated that the lowest-energy route between any two celestial bodies is an ellipse that forms a tangent to the starting and destination orbits of these bodies. Such a transfer orbit between the Earth and Mars is graphed in the following illustration. A spacecraft traveling from Earth to Mars along the Hohmann trajectory will arrive near Mars's orbit in approximately 18 months. Just a small application of thrust is all that is needed to put a space probe into a circular orbit around Mars. The Hohmann transfer applies to any two orbits, not just those with planets involved (see Figure 1).

In the figure, Hohmann Transfer Orbit (light gray oblong ring), Earth's orbit is represented by the white circle, and Mars' orbit is represented by the darker gray circle. A spaceship leaves from point 2 in Earth's orbit and arrives at point 3 in Mars's.

Another example of navigational technique is routinely called the "gravitational slingshot." It utilizes the gravitational influence of planets and their moons to change the speed and direction of a space probe without the application of an engine. In this case, a spacecraft is sent to a distant planet on a path that is much faster than the Hohmann transfer. This would typically mean that it would arrive at the planet's orbit and continue past it. However, if there is a planetary mass between the departure point and the target, it can be used to bend the path toward the target, and in many cases the overall travel time is greatly shortened. Prime examples of the gravitational slingshot are the flights of the two spacecraft of the

American Voyager program, which used slingshot effects to redirect trajectories several times in the outer solar system. Astrodynamics considers many other interesting approaches. Several technologies have been proposed that both save fuel and provide significantly faster travel than Hohmann transfers; most are still theoretical.

Because of astrodynamics limitations, travel to other solar systems bodies is practical only within certain time windows. Outside of such windows, these bodies are essentially inaccessible from Earth using current technology. Mathematicians helped design the Interplanetary Superhighway, a network of low-energy trajectories, in order to find efficient routes through space; these mathematical foundations originated with French mathematician Henri Poincaré.

Achievements and Obstacles

The modern accomplishments in interplanetary travels are extraordinary. Remotely guided space robots have flown past all of the planets of the solar system from Mercury to Neptune, and the National Aeronautics and Space Administration's (NASA's) spacecraft New

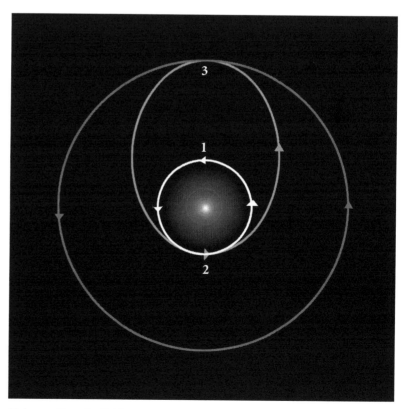

Hohmann Transfer Orbit.

Horizons is scheduled to fly past the dwarf planet Pluto in 2015. The five most distant spacecraft (including the American ships *Pioneer-10*, *Pioneer-11*, *Voyager-1*, and *Voyager-2*) were scheduled to leave the solar system at the beginning of the twenty-first century. Artificial satellites have orbited Venus, Mars, Jupiter, and Saturn. Spacecraft have landed on the Moon, Venus, Mars, Saturn's moon Titan, and asteroid 433 Eros. The first probes to comets (European *Giotto*, Russian *Vegas*, American *Stardust*) were fly-by missions. In 2005, the *Deep Impact* probe hit the comet 9P/Tempel to study the composition of its interior.

Great achievements took place in manned interplanetary travels once mathematicians, scientists, and engineers understood the mathematical principles required to launch spacecraft outside Earth's atmosphere and to maneuver in the microgravity environment of space. NASA also recruited astronauts with strong academic credentials in science and mathematics. America's Mercury and Gemini programs put humans into space and Earth orbit and taught them how to change trajectory in space to move to a new orbital altitude or to dock with other spacecraft, while the Apollo program took them to the moon. After missions in which men orbited the moon and returned, Apollo 11 landed astronauts Neil Armstrong and Edwin "Buzz" Aldrin on the moon in 1969. There were six successful manned American expeditions to the moon from 1969 to 1972.

Further development of interplanetary travel has many obstacles that will require a great deal of mathematical analysis to model, simulate, and solve. For example, astronauts must be protected from extreme radiation exposure in the Van Allen belt, a torus-shaped region of space surrounding the Earth and other planets named after geophysicist James Van Allen of Iowa.

The larger outer radiation belt is about four Earth radii (RE) above the surface of the Earth and the inner is about 1.6 RE, with a gap at roughly 2.2 RE. Apollo astronauts were briefly exposed to this radiation on trips to the moon. Conspiracy theorists who disputed the notion that humans landed on the moon cited the Van Allen belt as evidence that the astronauts would have died from radiation, but simple calculations and the data collected by radiation sensors worn by astronauts (similar to those worn by scientists and hospital workers who may be exposed to radiation) demonstrated that the speed and design of the Apollo capsules protected astronauts during these relatively short trips.

If the Earth was the main focus of many sciences (geodesy, geology, geophysics, geochemistry, and oceanography) for millennia, interplanetary travel created a new important branch of research—comparative planetology—which is essential for understanding the history of Earth and its evolution.

Among many other difficult problems of interplanetary travel is developing adequate human life support. A breathable atmosphere must be maintained, with adequate amounts of oxygen, nitrogen, controlled levels of carbon dioxide, trace gases, and water vapor. It is also necessary to solve the problem of food supply.

At some point in time, all of these problems may be overcome. Incentives for future expansion of interplanetary flights include the possibility of colonizing other portions of the solar system and utilizing resources.

Alexander A. Gurshtein

Further Reading

Battin, Richard. *An Introduction to the Mathematics and Methods of Astrodynamics*. New York: American Institute of Aeronautics and Astronautics, 1999.

Benson, Michael. *Beyond: Visions of the Interplanetary Probes*. New York: Harry N. Abrams, 2003.

Kemble, Stephen. *Interplanetary Mission Analysis and Design*. Berlin: Springer, 2006.

Launius, Roger D. *Frontiers of Space Exploration*. Westport, CT: Greenwood Press, 2004.

Launius, Roger D., and Howard E. McCurdy. *Robots in Space: Technology, Evolution, and Interplanetary Travel*. Baltimore, MD: Johns Hopkins University Press, 2008.

Zimmerman, Robert. *Leaving Earth: Space Stations, Rival Superpowers, and the Quest for Interplanetary Travel*. Washington, DC: J. Henry Press, 2003.

MARS: ILLUSIONS AND CONSPIRACIES

Categories: Mars; Planets and Planetology

Misinterpretations of telescopic observations led to the illusion of artificial canals on Mars, and visits of increasingly sophisticated spacecraft to the planet resulted in detailed photographs in which some enthusiasts saw faces and pyramids but space scientists saw buttes and mesas. These fanciful misperceptions can serve as a cautionary tale against the tendency by observers with strong preconceptions to see what they want to see rather than what is really there.

Overview

Some astronomers see the history of their discipline as the supersession of erroneous beliefs (or illusions) by true ideas that, over time, become ever more faithful descriptions of reality. In this way, the spherical Earth replaced a flat one, heliocentrism superseded geocentrism, and an expanding cosmos of multitudinous galaxies supplanted a static, small Milky Way universe. Outside Earth, Mars (the Romans' name for Ares, as it was known to the ancient Greeks) has been the subject of more illusions and conspiracies than any other planet. For centuries this "Red Planet" has fired the imagination of humankind, and the invention of the telescope and even visitations by orbiters and landers have not dampened the desires of many to see Mars as a locus of past or present intelligent life.

During the nineteenth century, ever more powerful telescopes allowed astronomers to see yellow clouds and white polar caps on the Martian surface. These astronomers commonly interpreted the polar caps, which expanded and diminished with the seasons, as frozen water. Others interpreted the dark areas as seas and the light areas as land. Pietro Angelo Secchi, a Jesuit astronomer who was the first to make multicolored drawings of Mars, was also the first to use the Italian term *canale* for a prominent feature (later called Syrtis Major). Another Italian astronomer, Giovanni Schiaparelli, created, starting in 1877, extremely detailed drawings of the Martian surface with hundreds of named features, including a growing number of rectilinear *canali*. Although *canale* may mean "canal," as in Canale di Panama, Schiaparelli interpreted these Martian *canali* as "channels" or "grooves," that is, as natural, not artificial, structures. He viewed these interlocking channels as a "hydrographic system" through which melting polar waters circulated throughout the planet, helping to foster organic life. He was "absolutely certain" that he had seen this intricate network, and he was even open to the possibility that the *canali* might be artificial, but he dismissed as imaginary the convoluted maps and interpretations of observers such as Percival Lowell.

For years the American businessman and amateur astronomer Lowell had successfully managed the family's great fortune, based on textiles, finance, and land, but, having read about Schiaparelli's *canali*, he became passionately interested in Mars and used his wealth to build an observatory near Flagstaff, Arizona, whose altitude and dry desert air facilitated Lowell's ability to see many more *canali* than Schiaparelli had detected. Lowell also became convinced that the complex system of hundreds of canals that he mapped was due to the constructive skills of intelligent Martians. He even invented a story to explain how these massive canals, which needed to be many kilometers wide to be visible from Earth, were built to bring polar water to the warm equatorial regions to support life on a planet growing colder and drier. He published his data, maps, and interpretations in three books, *Mars* (1895), *Mars and Its Canals* (1906), and *Mars as the Abode of Life* (1908), which were popular with the public, journalists, and science-fiction writers but were criticized by many astronomers and scientists. In the twentieth century, astronomers, who had the advantage of powerful refracting and reflecting telescopes, were unable to see either Schiaparellian or Lowellian canals. In 1907, Alfred Russel Wallace, who had discovered natural selection independently of Charles Darwin, published a book that attacked Lowell's hypothesis of a Mars inhabited by intelligent beings by showing that the planet

Mars Compared with Earth

Parameter	Mars	Earth
Mass (10^{24} kg)	0.64185	5.9742
Volume (10^{10} km³)	16.318	108.321
Equatorial radius (km)	3,396	6,378.1
Ellipticity (oblateness)	0.00648	0.00335
Mean density (kg/m³)	3,933	5,515
Surface gravity (m/s²)	3.71	9.80
Surface temperature (Celsius)	−128 to +24	−88 to +48
Satellites	2	1
Mean distance from Sun		
millions of km (miles)	228 (141)	150 (93)
Rotational period (hrs)	24.63	23.93
Orbital period	687 days	365.25 days

Source: National Space Science Data Center, NASA/Goddard Space Flight Center.

was so cold and dry that it was "absolutely uninhabitable." Although a few astronomers such as Earl C. Slipher came to Lowell's defense, many were skeptical, since they failed to find any water vapor in the Martian atmosphere and since those who saw the canals failed to agree on their locations or dimensions. Nevertheless, Slipher, who took 126,000 photographs of Mars at Lowell Observatory from 1906 to 1962, insisted that many of these pictures verified the existence of linear canals.

Illusions about Mars increased through publications such as Immanuel Velikovsky's *Worlds* in Collision (1950), in which he used astronomical data, biblical stories, and folk legends to argue that the Earth, Venus, and Mars had exchanged atmospheres when, only a few thousand years before, they nearly collided with each other. Even though the American scientific community vigorously attacked Velikovsky's ideas, which were easily falsified by scientific information and analysis, he developed a cult following. He and his followers attributed the derision of scientists to a conspiracy to cover up the real truth behind the solar system.

In 1965 the American spacecraft Mariner 4 became the first to send back pictures of the Martian surface from a relatively close range. These pictures confounded the believers in canals, Mars cultists, the public, and astronomers, who were surprised by an extensively cratered world that was more Moon-like than Earth-like. Furthermore, Mars's extremely thin atmosphere, only one two-hundredth as extensive as Earth's atmosphere, was unable to support any liquid water. Other Martian probes, such as Mariners 6 and 7, appeared to confirm the view of Mars as lifeless and dull, far from the exciting vision of Lowell and his followers.

Knowledge Gained

The conception of what some astronomers called the "old Mars" was transformed by Mariner 9, which arrived at the planet late in 1971 and comprehensively mapped its surface in 1972. Its dramatic images revealed a new and more interesting Mars, with such fascinating features as Olympus Mons, the

solar system's largest volcano, and Valles Marineris, a complex of ravines so vast that it could hold nearly five hundred of the Earth's Grand Canyon. Mariner 9's many detailed pictures of mountains, craters, mesas, plains, valleys, and other features provided areologists—those who study Mars—with an abundance of data to construct a geological history of Mars. These images also, however, encouraged speculators to develop controversial interpretations of ambiguous features. These speculations became highly imaginative when sharper pictures of Mars were returned to Earth by Viking Orbiter 1, especially of Cydonia, a plain with scattered buttes and mesas in the middle northern latitudes. One photograph, taken on July 25, 1976, showed a mesa that, in certain light, vaguely resembled a human face, and even though space scientists saw it as natural, similar to anthropomorphic geological features on Earth, others interpreted it as a remnant of a vanished civilization.

When, in 1979, computer scientists released enhanced images of the Cydonian region, some interpreters saw

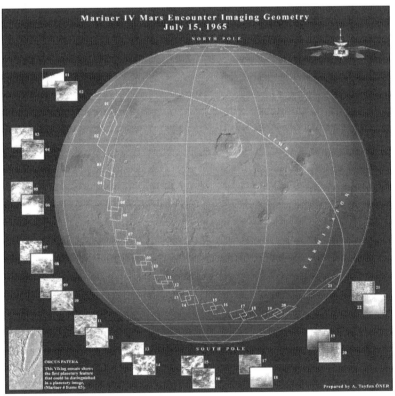

In 1965, Mariner 4 was the first probe to reach Mars. It supplied data and images that initiated our modern understanding of the Red Planet: a cratered, harsh surface rather than a lush environment capable of supporting life. (NASA)

such new features in the "face on Mars" as eyes with pupils, a nose, a mouth with teeth, and a headdress reminiscent of an Egyptian pharaoh's. These interpreters also claimed to see several pyramids in the region surrounding the "face." These pyramids were hundreds of times larger than the Great Pyramids at Giza in Egypt, and, according to these interpreters, their geometrical shapes, which were a clever combination of pentagonal and hexagonal symmetries, as well as their architectural alignments, manifested a knowledge of astronomical and mathematical constants.

Magazine articles, books, and Web sites soon proliferated, advancing the hypothesis that the Cydonian region contained the remains of a colossal city. Richard Hoagland, a journalist who had become interested in the Cydonian structures in 1980, published a book, *The Monuments of Mars* (1987), which became so popular that it went through four editions in the late 1980's and into the 1990's. When space scientists ridiculed the interpretations of the "Cydonian cultists" by showing that their "pyramids" were actually natural structures that had been sculpted by wind and water, not intelligent beings, the "Cydonians" responded by founding such organizations as the Academy of Future Science and the Independent Mars Investigation Project to defend their views. Using publications, Web sites, radio and television appearances, and DVDs, these Cydonians alleged that a conspiracy of the National Aeronautics and Space Administration (NASA) and other governmental agencies had developed to cover up the evidence of intelligent life on Mars, since this explanation challenged established social, political, and religious values. When such missions as the Mars Observer failed, some Cydonians accused NASA scientists of deliberately sabotaging the mission to prevent the public from learning "the truth" about the "face and pyramids." When such NASA missions as the Mars Global Surveyor, Mars Pathfinder, and Odyssey succeeded, returning data that showed the "face" as an eroded conical butte and the "pyramids" as irregularly shaped mesas, the Cydonians reprimanded the scientists for interpreting these new photographs through their academic prejudices. Some critics wondered whether the evidence gathered by a lander or astronauts in the region of the supposed face and pyramids would be sufficient to resolve the conflicts between space scientists and the Cydonians.

Context

Throughout history, genuine scientists have often contended with pseudoscientists. Sometimes, a pseudoscience has evolved into a genuine science (as in the case of alchemy, which became chemistry), but astronomers have been unable to dispatch pseudoastronomies such as astrology. As planetary scientist Carl Sagan often remarked, even in such advanced societies as the United States, more professional astrologers were making a living than professional astronomers. Many space scientists considered any attention paid to pseudoscientific theories about Mars as a waste of their time and taxpayers' money. For example, instead of spending valuable time, money, and human energy in fulfilling the assigned mission of the Mars Global Surveyor, scientists, because of pressure from the media, public, and politicians, were forced to divert the spacecraft to take pictures of the "face on Mars" and other supposed "alien constructs." These new pictures clearly revealed, in the view of the scientists, that the objects in question were naturfacts, not artifacts.

Other scholars have studied these illusions and conspiracies for the insights that they provide into fascinating psychological and sociological phenomena. For example, the cases of Schiaparelli and Lowell are illustrative of "pathological science," in which scientists erroneously convince themselves of the genuineness of what they are seeing and saying. Pathological science often occurs when threshold effects are involved—for example, when scientists attempt to interpret things that are barely visible. In these cases humans tend to overinterpret scattered dark features into a line, as Schiaparelli and Lowell did. In the case of the Cydonian cultists, other forms of misinterpretation have surfaced. For example, modern computerized image enhancement can be overdone, sometimes leading to the appearance of features that were not present in the original. Sociological pressures often help explain how Cydonians see themselves as the victims of various conspiracies. They find pleasure and comfort in the idea that intelligent creatures once inhabited Mars, but these extraordinary claims require extraordinary proofs. However, scientists, relying on data collected by superior orbiters and landers, have seen nothing that cannot be better understood as the consequence of natural forces. History has shown that humans can fool themselves, but scientists, based on long experience, have found that "Mother Nature" cannot be fooled.

Robert J. Paradowski

Further Reading

Goldsmith, Donald, ed. *Scientists Confront Velikovsky.* Ithaca, N.Y.: Cornell University Press, 1977. Eminent scientists such as Carl Sagan show why, from a detailed study of astronomical evidence, Velikovsky's theories about the solar system in general, and Mars in particular, must be regarded as pseudoscience. Bibliography, index, and references for each chapter.

Hartmann, William K. *A Traveler's Guide to Mars: The Mysterious Landscape of the Red Planet*. New York: Workman, 2003. Called a "masterpiece of scientific writing for the general reader," this lavishly illustrated book provides a fascinating survey of the "New Mars," including sections and sidebars on how astronomers have responded to illusions about Mars. Glossary, selected sources, and index.

Hoyt, William G. *Lowell and Mars*. Tucson: University of Arizona Press, 1976. A critical but balanced portrait of Lowell, his observatory, his writings, and the influence of his ideas on the public, with an account of the discovery of Pluto, named in his honor.

Kargel, Jeffrey S. *Mars: A Warmer, Wetter Planet*. New York: Springer Praxis, 2004. A detailed analysis of the new face of Mars that has been revealed by the evidence gathered by orbiters and landers. Illustrated with a color section and many black-and-white photographs. Bibliography and index.

Kieffer, Hugh H., et al., eds. *Mars*. Tucson: University of Arizona Press, 1992. This massive compendium, part of the Space Science series, is the product of 114 collaborating experts. Part 1 deals with the history of explorations of Mars visually, telescopically, and through spacecraft. Here the subject of illusions about Mars is also treated. Has an excellent section, "Books About Mars," on pages 19–24. Illustrated with photographs and diagrams. Includes a glossary, a bibliography, and an index.

Sheehan, William, and Stephen James O'Meara. *Mars: The Lure of the Red Planet*. Amherst, N.Y.: Prometheus Books, 2001. Sheehan, a psychiatrist and amateur astronomer, analyzes the history and controversies connected with human encounters with Mars through the naked eye, telescopes, and modern spacecraft. Intended for general audiences, the book has appendixes offering data on Mars, notes to original and secondary sources, and an index.

Wilford, John Noble. *Mars Beckons: The Mysteries, the Challenges, the Expectations of Our Next Great Adventure in Space*. New York: Alfred Knopf, 1990. Wilford, a two-time Pulitzer Prize winner, discusses the history of both illusions about and hard-won knowledge of Mars, with a concluding treatment of future plans for getting humans on the planet. Appendixes on the Mars missions and on the solar system. Bibliography and index.

MARS: POSSIBLE LIFE

Categories: Life in the Solar System; Mars; Planets and Planetology

Environmental conditions in the past suited the origin of life on Mars. A primary motive for sending orbiters and landers to the planet has been to detect Martian life; it would be a principal goal for a crewed mission as well. If detected, such life would help elucidate the origin of life on Earth and, possibly, elsewhere in the universe. If not detected, the question of whether terrestrial life is unique would remain open.

Overview

On the basis of their observations from telescopes, photographs and remote sensors from orbiters, and direct inspection by means of experiments performed by landers and rovers on the planet's surface, scientists have ruled out the presence of intelligent life on Mars, fourth planet from the Sun and Earth's immediate outer neighbor. No direct evidence of life of any kind has been found. However, circumstantial evidence that life might have existed in the past, and could persist even today, steadily accumulated after the first probe photographed Mars in 1964. In all cases, it is evidence based on comparisons with the conditions that support life on Earth.

Mars specialists base their search on the broad biological definition of life as a chemical system capable of Darwinian evolution to accommodate to changing environmental conditions. Most agree that this entails the ability to process energy and nutrients, grow, and reproduce. Accordingly, the search is on for environments on Mars that could foster these activities. Unfortunately, it is a harsh world. The Viking landers of the mid-1970's confirmed what scientists had suspected, that iron oxide, poisonous to life on Earth, permeates the planet's surface. The atmosphere is thin, frigid, dry, and very low in oxygen. The absence of a planetary magnetic field permits intense ultraviolet radiation, also deadly to life, to reach the surface. These facts suggest that the existence of life above ground is highly unlikely.

Water is an essential element for life, although its presence does not guarantee life. According to evidence from orbiters and landers, Mars had oceans, lakes, rivers, and a thicker atmosphere in the past, and may still see occasional outbursts of underground water on the surface. However, most water now exists as ice in polar ice fields and as permafrost beneath the surface, a region known as the cryosphere. Below these there may be a hydrosphere, a band of water-permeated rock. Most scientists believe that life could thrive at the boundary between the cryosphere and hydrosphere. There is some evidence to encourage that belief. Probes have detected methane in Mars's atmosphere; methane is a waste gas from biological processes, and it is known as a biomarker or biosignature. Because methane lasts in the atmosphere for only a few hundred years, it must be replenished. That source could be Martian organisms now alive, although theoretical inorganic chemical processes also have been proposed as the source.

If organisms exist in the cryosphere-hydrosphere boundary region or as spores near ancient water bodies, they are probably not large. There could be multicellular organisms like the tube worms that feed from hydrothermal vents in the cold dark waters of the deep oceans on Earth, but most scientists foresee finding only single-cell organisms. Terrestrial organisms known as extremophiles live in conditions ranging from 253 to 394 kelvins (−4° to 250° Fahrenheit), in a wide range of acidity, in very salty water, without light, or kilometers underground. Some extract energy and nutrients from hydrothermal vents (hyperthermophiles), such as hot springs. Others feed from inorganic chemicals in rocks (chemolithoautotrophs)—for instance, ingesting sulfide minerals and excreting sulfuric acid—and in slushy water or salty water colder than the normal freezing point (psychrophiles). It is Martian equivalents of such extremophiles that scientists hope to find.

In a subsurface ecosystem Martian organisms would show variety in form and function. The basic structure ought to be a cell with a semipermeable membrane, such as the sack of lipoproteins defining most terrestrial cells, and internal structures that split apart chemicals in order to use the by-products in their metabolism. Most would be grazers, feeding off the ambient nutrient source, but there are likely to be predators that consume the grazers. Either may have a means of locomotion, such as cilia or the ability to expand and contract. In order to evolve, they would need some type of chemical record of their mechanism, as exists in the ribonucleic acid (RNA) and

deoxyribonucleic acid (DNA) of Earth organisms, to pass on during reproduction. Because the temperature and air pressure on Mars are low, the metabolism, reproduction, and movement of organisms may well be sluggish and sparse in comparison to those functions in Earth organisms. They could live from as little as 35 centimeters (1.1 foot) below the surface to as much as 10 kilometers (6.2 miles), scientists speculate.

Martian life may have once thrived and then declined and finally vanished as the ancient surface water dried up. In that case, fossils might remain. Paleobiologists have found fossilized forms of Earth's single-cell organisms, as in stromatolites, that date back at least 3.5 billion years before the present. Should such fossils survive on Mars, they would establish that life rose there and provide clues about which environmental conditions supported an ecosystem. In 1996 scientists from the National Aeronautics and Space Administration (NASA) announced finding evidence for life in a Martian meteorite known as ALH 84001, discovered in Antarctica. Among the evidence was what appeared to be a fossilized cellular structure. Although the majority of scientists later discounted the evidence and judged the microscopic structure to be the result of nonbiological processes, research demonstrates that the conditions for preserving delicate structures, such as fossils, exist on Mars. Additionally, it is possible that early conditions were favorable for life but then, because of meteorite bombardments 3.9 billion years ago, changed before life actually got started. In that case, scientists might still detect the precursors to life in complex organic chemicals.

Two further possibilities for life on Mars worry scientists. Just as ALH 84001 was blasted off Mars by a meteorite and made its long journey to Earth, a chunks of Earth probably reached Mars, and they could have carried Earth organisms there, seeding the Red Planet with terrestrial life. It is probable, moreover, that the landers sent to Mars by American, Russian, and European space agencies carried organisms with them despite decontamination protocols. Such "forward contamination" would be even more likely from a crewed mission to the planet. Seeding and contamination do not mean that Earth organisms have survived in the harsh Martian environment. On the other hand, if Earth organisms have adapted, scientists may have difficulty establishing their origin.

Knowledge Gained

Experiments performed aboard the Viking landers (1976) established that Mars's soil contains chemicals inimical to life, particularly oxidants such as iron oxide (rust), and the Sojourner, Opportunity, and Spirit rovers confirmed the finding at other locations. These findings do not absolutely rule out life. Viking experiments may have missed a biomarker that they were not designed to detect, or the rocket exhaust from their landing may have killed organisms within their reach. In any case, the search for life taught scientists much about the chemical nature of the soil and the distributions of oxidants.

In the 1990's NASA set a policy for its Mars probes: "Follow the water." The resulting search from orbit and on the surface found evidence of erosion and chemical deposits, such as hematite, that on Earth derive from flowing water. Sensors in orbiters detected underground ice as well. These discoveries not only encourage scientists to look further for life but also suggest that sufficient water exists to support human habitation on the planet. At the same time, the various landers and rovers, all of which far exceeded their expected performance, proved the versatility and hardiness of technology in the Martian environment.

When it became probable that no life inhabited the Martian surface, scientists investigated the possibility of organisms living below the surface. The research depended on analogy to similar Earth habitats, and scientists were inspired to search for life in heretofore unexplored realms. The result was to expand greatly the knowledge about terrestrial extremophiles near volcanic vents deep in the oceans, in gelid water, in porous underground rock, or in salty, acidic, or alkaline conditions. Paleobiologists uncovered fossilized organisms from much further in the past than previously suspected.

In late 2008, analysis of data from NASA's Mars Reconnaissance Orbiter (MRO)—reported just after that spacecraft was given a two-year extension beyond completion of its primary mission—indicated that Mars in its distant past must have had sufficient water flowing across the surface that clay-rich (carbonate) minerals could form. MRO picked up the signatures of those clay minerals. That Mars may have been wetter and favorable for primitive life to develop is evidenced by the fact that these clay minerals survived to the present day. As Mars began losing its water, becoming a drier planet, the remaining water would have become acidic. Carbonate clay minerals are relatively easily dissolved in acidic water. Areas where the clay minerals survived on the surface to the present era would have been locations less hostile to life as the planet continued to become drier and drier. Although highly suggestive, this new information does not actually provide direct evidence that life may ever have existed on Mars.

Another discovery, this one made by ground-based telescopes atop Mauna Kea, Hawaii, was reported in an early January, 2009, *Science Update* aired on NASA Television. Reporting scientists explained that the infrared signature of methane gas had been detected in significant amounts in the Martian atmosphere. This held the potential for indirect evidence that primitive life might exist on the Red Planet in the present era. Since solar ultraviolet radiation penetrates through the thin Martian atmosphere, methane molecules dissociate, and over the eons of geologic time the methane should have dissipated without replenishment. Replenishment was possible by either one of two mechanisms: a geological mechanism involving the conversion of iron oxide to serpentine minerals in the presence of water, carbon dioxide, and a heat source, or a biological mechanism involving digestion occurring in primitive microorganisms. More direct examination of methane vents on the planet's surface would be needed to determine if methane production is of geological or biological origin.

Negative findings are also important, as much to human culture as to science: Mars supports no civilization or, in all probability, animal life. The alteration in surface color through the Martian seasons that fascinated astronomers like Percival Lowell comes from wind storms, not vegetation. Thereby, scientists discounted the possibility, so popular in science fiction, that Martian life poses a threat to Earthlings.

Context

The absence of Martians as imagined by H. G. Wells, Robert A. Heinlein, or Ray Bradbury has not dampened popular enthusiasm for the search for life on Mars. It endures because it promises to answer a question that has long made humanity look to the heavens and wonder: "Are we alone?"

The discovery of life on Mars would have profound implications. If proven to be entirely independent of life on Earth, that finding would mark a shift in understanding the universe as great as that of the Copernican Revolution. Philosophy would be tasked to reconsider the human moral obligation to other organisms. Religions would have to cope with the fact that life on Earth is not a unique creation. Science would be encouraged to look for life on still more worlds, such as the satellites of Jupiter, and for biomarkers in the light from other planetary systems. At the same time, NASA and other space agencies would need to take measures to protect Martian life from Earth organisms hitchhiking aboard planetary probes, or, in the case of a sample-return mission or crewed mission, to protect Earth's ecosystem from Martian organisms.

If Martian organisms were found to be related to those on Earth, the knowledge would also be fundamentally important. It would establish the great durability of life in spreading from one planet to another and leave open the possibility, as proposed in the Panspermia hypothesis, that both Mars and Earth were seeded long ago with life that originated elsewhere.

Proof that life never existed on Mars would be significant as well. The question of Earth's uniqueness would remain unsettled, yet scientists would learn an important fact: The types of chemical and geophysical conditions found on Mars are not conducive to life. Why that should be true would pose a major question for further research.

Roger Smith

Further Reading

Elkins-Tanton, Linda T. *Mars*. New York: Chelsea House, 2006. In a style suitable for high school students, the author, a geophysicist, provides a technically specific overview of Mars's environment and geologic history, including a short, reasonable summary of the case for life. With glossary, appendixes on basic relevant science, and photos.

Forget, Françoise, Françoise Costard, and Philippe Lognonné. *Planet Mars: Story of Another World*. Chichester, England: Praxis, 2008. The lucid text, suitable for high school students, explains current knowledge about the geology, hydrology, and meteorology of Mars. Three short chapters discuss possible life, and a fourth summarizes knowledge gained by space probes. With abundant photographs and graphics.

Hartmann, William K. *A Traveler's Guide to Mars: The Mysterious Landscapes of the Red Planet*. New York: Workman, 2003. This entertaining overview of the Red Planet includes a section about the physical conditions on Mars that could support life, and the text is a model of popular exposition. With many photographs and illustrations by the author, a planetary scientist, writer, and artist.

Kargel, Jeffrey S. *Mars: A Warmer, Wetter Planet*. New York: Springer Praxis, 2004. A planetary scientist, Kargel provides a comprehensive, sophisticated survey of the geology and climate of Mars throughout its history. The importance of water is a central theme, and he discusses the possibilities for life. With many graphics and photographs.

Kiang, Nancy Y. "The Color of Plants on Other Worlds." *Scientific American* 298 (April, 2008): 48-55. Although not specifically addressing possible Mars organisms, this article explains the nature and type of chemical evidence that scientists think will signal the presence of life in alien environments.

Walter, Malcolm. *The Search for Life on Mars*. Cambridge, Mass.: Perseus Books, 1999. Charming, concise, and equitable, this volume is especially valuable for Walter's explanations of how terrestrial paleobiology provides clues of what life on Mars might be like and the controversies and personalities behind the research.

MARS'S ATMOSPHERE

Categories: Mars; Planets and Planetology

Because many similarities exist among the planetary atmospheres, the study of one planet may contribute to the understanding of others. The atmosphere of Mars, with its simple structure, can be used to model certain aspects of Earth's atmosphere and is therefore a valuable aid in comprehending the past, present, and future of that atmosphere.

Overview

One goal of planetary scientists is to unravel the history of the Earth in the context of the origin and evolution of the solar system. Because all data record what is presently observed, conditions and processes of the past must be inferred from current conditions and processes. Scientists construct a model of an atmosphere's history from those inferences and use that model to project the future evolution of the atmosphere. The model is revised as better data are gathered.

Astronomers generally assume that the terrestrial planets (Mercury, Venus, Earth, and Mars) have had two atmospheres. Primitive atmospheres would have existed soon after the formation of the planets and would have been distinctly different in composition from those that exist today. These atmospheres contained hydrogen, helium, and other lightweight compounds with speeds near or above the escape velocities of terrestrial planets.

Consequently, primitive atmospheres would be lost over a reasonably short time, even as the planets cooled.

Secondary atmospheres developed as nitrogen, carbon dioxide, water, and argon were released from the planetary interiors as molten rock outgassed and volcanoes erupted. Volcanoes on Earth emit mostly carbon dioxide and water. Because scientists assume that the planets formed from the same cloud of protoplanetary material, they also assume that the gases coming from the interiors of different terrestrial planets would be similar. Therefore, the secondary atmospheres of Venus, Earth, and Mars should contain the same compounds, but the specific quantities of those compounds should vary as conditions on the planets vary. Venus should have mostly water vapor because it is nearer the Sun and has a higher surface temperature than does Earth. Mars should have mostly solid water because its distance from the Sun makes it a colder planet than Earth. Of course Earth has water in all three phases: gas, liquid, and solid.

Until the first spacecraft reached Mars, there was no accurate method for determining the pressure, composition, and temperature profile of its atmosphere. Mars was expected to have a thin atmosphere, with its low escape velocity of 5.0 kilometers per second. Most estimates of Mars's atmospheric pressure made during the first half of the twentieth century were between 1.2 and 1.8 pounds per square inch (Earth's is 14.6 pounds per square inch at sea level)—high enough for liquid water to exist on the surface of Mars as long as the temperature is not above 310 kelvins. Most scientists did not believe that Mars could retain much water, but if pressure estimates were accurate, the existence of water on Mars could not be eliminated from models of the planet and its atmosphere.

Composition of the Martian atmosphere was difficult to determine from Earth because the Earth's own atmosphere obscured much of the information that came from the planet. Because the atmospheres of the terrestrial planets were expected to have similarities, astronomers looked for nitrogen, oxygen, water, and carbon dioxide. Nitrogen is very difficult to observe, so no one was surprised when it was not detected in visible light. Astronomers, however, expected to find a significant amount of nitrogen when a spacecraft arrived at the planet. Earthbound telescopes also failed to detect oxygen or water but found carbon dioxide. Astronomers used this information and their assumptions to predict that the Martian atmosphere was largely nitrogen with some carbon dioxide present.

During the late 1960's, a series of United States Mariner spacecraft flew past Mars and found that atmospheric pressure was less than 0.09 pound per square inch. Roughly 95 percent of the atmosphere was carbon dioxide, with 1-3 percent nitrogen. Many astronomers concluded that Mars was dead and Moon-like. More recent analysis of the Martian atmosphere has shown its composition to be 95.3 percent carbon dioxide, 2.7 percent nitrogen, 0.13 percent oxygen, 0.03 percent water, and 1.6 percent argon, with trace amounts of krypton and xenon.

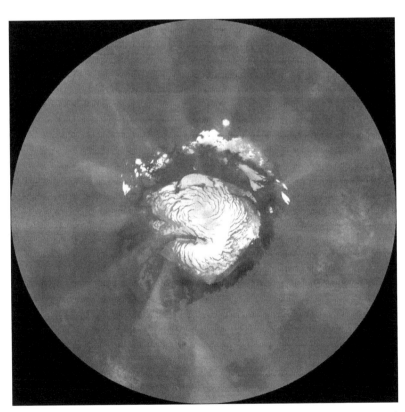

On October 11, 2008, the Mars Reconnaissance Orbiter's Mars Color Imager took this picture of the north polar ice cap on Mars; a 37,000-kilometer-diameter dust storm is moving counterclockwise through the Mars Phoenix landing site. (NASA/ JPL-Caltech/Malin Space Science Systems)

Average temperatures at the surface of Mars are about 215 kelvins. The warmest spot on the planet may reach 300 kelvins near local noon, but temperatures drop to less than 192 kelvins at the Viking 2 landing site. Polar regions are much colder, as confirmed in 2008 by the Mars Phoenix lander after it touched down on the Red Planet's northern polar region. Liquid water will not exist under these conditions. During the summer days, however, the surface temperature is high enough for liquid water to exist briefly. Because the vapor pressure of water in the atmosphere is very low, this liquid water evaporates quickly. As winter begins, the water molecules freeze to cold dust particles in the atmosphere. Carbon dioxide molecules also attach themselves during the cold nights, and when particles have enough mass, they fall to the surface. The temperatures during the day are high enough to vaporize carbon dioxide but not water. Pictures of the soil around the Viking 2 lander show a frost of these water-ice-coated dust grains. Clouds or ground fog of water ice crystals form about half an hour after dawn in areas heated by the Sun. Beyond latitudes of 65°, winter conditions cause carbon dioxide to freeze, and a hood of carbon dioxide clouds and haze hangs over the polar regions. The coldest regions on Mars are the poles during winter. Polar caps made of carbon dioxide ice change size as the seasons change, but a permanent cap of water ice remains at each pole, where the temperatures never get warm enough to allow water to melt or vaporize. Mars Phoenix dug trenches into the soil near the lander and found evidence of the permafrost layer not very far beneath the surface. The permanent cap at the south pole is smaller than the one at the north pole because Mars is much nearer the Sun during the south pole's summer than during summer in the northern hemisphere. The total atmospheric pressure changes by 26 percent seasonally because of the vaporization/condensation cycle of carbon dioxide at the poles. Much of the carbon dioxide vaporized at one pole moves toward the other pole and precipitates there.

In the northern and southern hemispheres, high-pressure systems form during summer months and low pressure develops during winter. The pressure difference is the greatest when it is summer in the southern hemisphere and winter in the northern hemisphere. Large-scale wind currents flow toward the north pole during northern winter and toward the south pole during southern winter. Dust particles picked up by these winds cause large-scale dust storms. These storms cause little erosion, because the thin Martian atmosphere can carry only small dust particles. Viking 1 measured wind gusts up to 26 meters per second as a dust storm arrived. The most vigorous storms may involve wind velocities greater than 50 meters per second. The Mars Pathfinder landing stage provided wind speed and direction data during 1997. It picked up mostly light winds, with direction varying considerably over the mission. The Mars Exploration Rovers imaged dust devils. Indeed some of the dust devils performed a useful function for the rovers as they blew off

Mars's Atmosphere Compared with Earth's

	Mars	Earth
Surface pressure	6.1-9 millibars	1.014 bars
Surface density (kg/m³)	~0.020	1.217
Avg. temperature (kelvins)	~215	288
Scale height (km)	11.1	8.5
Wind speeds (m/sec)	2-30	up to 100
Composition		
Argon	1.6%	9,430 ppm
Carbon dioxide	95.32%	350 ppm
Helium	—	5.24 ppm
Hydrogen	—	0.55 ppm
Hydrogen deuteride	0.85 ppm	—
Krypton	0.3 ppm	1.14 ppm
Methane	1.7 ppm	—
Neon	2.5 ppm	18.18 ppm
Nitrogen	2.7%	78.084%
Nitrogen oxide	100 ppm	—
Oxygen	0.13%	20.946
Water	210 ppm	1%
Xenon	—	0.08 ppm

Note: Composition: % = percentages; ppm = parts per million
Source: Data are from the National Space Science Data Center, NASA/Goddard Space Flight Center.

dust that had accumulated on solar panels. This cleaning effect boosted spacecraft power.

Some data imply that the Martian atmosphere was denser at one time than is presently observed. Channels (not canals) found by the Mariner spacecraft have the same appearance as channels formed on Earth by flowing water. Many craters show more erosion than is possible with the current atmosphere. Indirect evidence indicates that significant quantities of nitrogen, water, and carbon dioxide have been outgassed from the Martian interior. Many astronomers thus believe that the Martian atmosphere was once denser and warmer than now and that it became moist periodically as polar caps melted. Rivers may have flowed during these periods. As time passed, pressure gradually decreased as water vapor and carbon dioxide were lost from the atmosphere. Those losses reduced the capacity of the atmosphere to retain heat. A change in the tilt of Mars's rotational axis and a change in Mars's orbital path also reduced the temperature. As temperatures dropped, water from the atmosphere was permanently trapped in the polar caps. Eventually, much carbon dioxide was also deposited in the polar caps, and the current cycle of vaporization and condensation was established. Exposed solid water ice was seen to sublimate away during Mars Phoenix investigations in 2008.

In mid-January, 2009, the National Aeronautics and Space Administration (NASA) announced a discovery that held the potential to become a paradigm shift in the nature of Martian exploration by future robotic spacecraft. Using ground-based telescopes at great altitude atop Hawaii's Mauna Kea, astronomers detected the infrared signature of methane in the thin Martian atmosphere. Lacking protection from solar ultraviolet radiation, methane molecules in the atmosphere would be subject to dissociation. Thus, the detection of significant amounts of methane indicated that the gas was almost certainly being replenished. Methane production could have a geological origin, but it also could have a biological one. A geological source for Martian methane could involve the conversion of iron oxide into serpentine minerals, a process that would require water, carbon dioxide, and an internal heat source; this process does occur on Earth. A biological origin for Martian methane seen in the atmosphere would involve digestion processes inside microorganisms. The methane data NASA reported could not be used to distinguish between a biological or geological source for Martian methane production. As a result, the space agency began to consider sending the next planned rover, Mars Science Laboratory, to the location of a methane vent to conduct analyses that might determine whether biology was involved in methane production on the Red Planet or if it came from a geological process.

Methods of Study

Astronomers face the challenge of collecting data on objects that are millions of miles away. Earth-based telescopes collect light that is analyzed for relevant information. This technique, however, often does not provide the precision needed for study of planets. The advent of

At the Mars Phoenix Lander's site in the north polar region, a dust devil can be seen near the horizon (just right of the mast of the lander's meteorological station). (NASA/JPL-Caltech/University of Arizona/Texas A&M University)

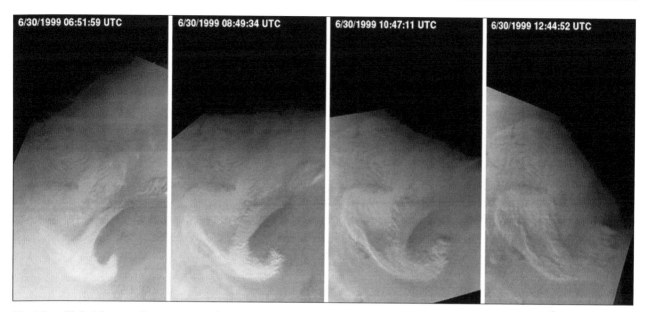

The Mars Global Surveyor's cameras sent back these images of a storm over Mars's north pole in 1999 as the Martian summer was ending and autumn beginning. (NASA/JPL/Malin Space Science Systems)

the space program gave astronomers new opportunities to gather data as sophisticated spacecraft traveled to the remote parts of the solar system. Mariner 4 took the first close-up spacecraft photos of Mars on July 15, 1965. It was followed by Mariner 6 and 7 flybys in 1969, the Mariner 9 Orbiter in 1971, and the Viking Orbiters and Landers in 1976.

Although the Mariner spacecraft had determined that carbon dioxide was the main component of the Martian atmosphere, concentrations of other compounds present in small amounts were still unknown. As the Viking landers descended through the atmosphere, they looked for nitrogen, argon, and other elements whose molecular weight was less than 50. A mass spectrometer, an instrument specifically designed to find compounds and identify them according to their masses, analyzed the atmosphere at altitudes above 100 kilometers. Nitrogen was discovered with an abundance of roughly 3 percent. Argon, an inert gas, was found with an abundance of 1.6 percent. Another instrument, the retarding potential analyzer, showed that many hydrogen and oxygen ions were escaping from the atmosphere. Because these two elements combine to form water, their loss can be expressed in terms of water loss. Roughly 240,000 liters of water were lost each day.

The gas chromatograph mass spectrometer was used in analyzing gases emitted from experiments

with the soil samples; however, it was also used to find krypton and xenon, which could not be detected by the upper atmosphere mass spectrometer because the atmospheric concentrations of these two elements were too low. Concentrations of these gases were enriched by a procedure in which Martian air was pumped into the sample chamber, from which carbon dioxide and carbon monoxide were removed. By repeating this process several times, krypton and xenon concentrations were increased to measurable levels.

The meteorology boom contained sensors to measure temperature, wind direction, and wind velocity. These sensors are thin bimetallic wires. As they change temperature, they induce an electric current. The size of this current determines the temperature. Wind cools the wires; by measuring the current passed through the wires to warm them to their original temperature, researchers can determine the velocity of the wind. Wind direction is measured by monitoring which of the sensors is cooling faster. Atmospheric pressure was measured by a sensor containing a thin metal diaphragm. The amount of movement of the diaphragm was used to determine the pressure.

Context

Planetary exploration has as its broad goals the search for life and for clues concerning the origin of the solar system. Atmospheric conditions that include the presence of water and nitrogen must exist in order for the basic materials for life to exist. The dense, hostile atmosphere of Venus destroyed any hope of finding life there. The Martian atmosphere in some ways resembles Venus's atmosphere more than Earth's; in fact, Venus and Mars each have about 95 percent carbon dioxide and 3 percent nitrogen, but atmospheric pressure on Venus is more than nine thousand times that on Mars. Although there is evidence that water once flowed on the Martian surface and that the atmosphere was once thicker than it is now, there is no absolutely conclusive evidence that there has been enough water in the atmosphere to cause rain. Conditions for the development of life do not seem to have existed in the Martian atmosphere, but much more investigation is necessary to discount the possibility.

Further exploration of the planet will continue to be conducted by robotic spacecraft. Astronomers would also like to send scientists to explore Mars and set up a long-term research base much like those international scientists have established at the South Pole of Earth. Because the Martian atmosphere has a simple structure and because there are no bodies of water to affect the airflow, the atmospheric movement follows predictable patterns. Study of this simple system could lead to a better understanding of the more complex atmospheres of Earth and perhaps Venus. On Earth, this understanding could lead to better prediction of weather patterns, increased agricultural production, and decreased danger from natural disasters. Because carbon dioxide plays an important role in the greenhouse effect, study of the Martian atmosphere could reveal important information on how to deal with the increasing concentration of carbon dioxide in Earth's atmosphere.

The beginning of the twenty-first century saw a tremendous concern that Earth might be suffering global warming, whether from natural causes or from human impact on the planet. However, understanding of atmospheric physics was insufficient to predict with any certainty the future course of changes in Earth's climate. A more basic issue than global warming needs to be investigated before a better model of Earth's total climate system can be developed: the conditions that led to the evolution of the terrestrial planets' atmospheres. Venus, Earth, and Mars have a great deal in common. Essentially they started with similar early conditions. Why did Venus develop a thick carbon dioxide atmosphere and a runaway greenhouse effect, which left the planet's surface tremendously hot, whereas Mars became arid and cold, with a thin carbon dioxide atmosphere? Why did Earth's atmosphere develop in a way that led to the production of life as we know it?

Astronomers also believe that a better understanding of Earth's atmosphere will help them to draw conclusions about the possible existence of other planets like Earth in other solar systems. If life can form from inorganic matter, a careful study of the atmospheres on Earth and Mars could set limits on the range of conditions suitable for life to exist. Such a study may even conclude that life cannot spontaneously erupt from nonliving matter. Such a conclusion would require a total revamping of modern scientific thought.

The major question to be answered in understanding the climate and atmospheric evolution on Mars involves where its water went. If that water remains on the planet in large quantities, it could have huge impact upon future exploration of the planet.

Dennis R. Flentge

Further Reading

Carr, Michael H. *The Surface of Mars.* Cambridge, England: Cambridge University Press, 2007. The author provides a complete description of the geological heating of Mars as understood based on results from Mars Global Surveyor, Mars Odyssey, Mars Reconnaissance Orbiter, Mars Express, and the Mars Pathfinder and Mars Exploration Rovers. Heavily illustrated with comprehensive reference list.

Consolmagno, Guy, and Martha Schaefer. *Worlds Apart: A Textbook in Planetary Sciences.* Englewood Cliffs, N.J.: Prentice Hall, 1994. Using low-level mathematics but integral calculus where required, this text demonstrates how the discipline of planetary science progresses by questioning previous understandings in light of new observations. Aimed at college students, but accessible to nonspecialists as well.

Encrenaz, Thérèse, et al. *The Solar System.* New York: Springer, 2004. A thorough exploration of the solar system from early telescopic observations through the space missions that had investigated all the planets by the publication date. Takes an astrophysical approach to give our solar system a wider context as just one member of similar systems throughout the universe.

Hartmann, William K. *Moons and Planets*. 5th ed. Belmont, Calif.: Thomson Brooks/Cole, 2005. An updated version of a classic text on planetary science written from a comparative planetology perspective rather than devoting individual chapters to each planet and body in the solar system. Material on Mars covers all aspects of both Earth-based and spacecraft observations of Mars.

Kargel, Jeffrey S. *Mars: A Warmer, Wetter Planet*. New York: Springer Praxis, 2004. A member of Springer Praxis's excellent Space Exploration series. The book takes the reader on a search for Mars's water. The author provides a convincing case that the picture of a dry, waterless world portrayed initially by the early Mariner probes is not the Mars of today's understanding.

Lewis, John S. *Physics and Chemistry of the Solar System*. 2d ed. San Diego, Calif.: Academic Press, 2004. Suitable for an undergraduate course in planetary atmospheres, but accessible to the general reader with a technical background.

Moore, Patrick. *Guide to Mars*. New York: W. W. Norton, 1977. Dated but useful for its overview of the historical development of information about the planet. Describes the pre-Mariner data through the Viking missions. Some black-and-white photos; several maps. For a general audience.

Morrison, David, and Tobias Owen. *The Planetary System*. 3d ed. San Francisco: Pearson/Addison-Wesley, 2003. A discussion of data from each of the planets. Contains a large number of photographs and line drawings. Although intended as an astronomy textbook, it provides good reading for anyone with an interest in the solar system. Some sections require a science background.

Zubrin, Robert. *Entering Space: Creating a Spacefaring Civilization*. New York: Tarcher, 2000. The author displays a gung-ho attitude toward making humanity a truly spacefaring species by accepting the challenge of journeying to Mars sooner rather than later. Speculates beyond travel to Mars.

MARS'S CRATERS

Categories: Mars; Planets and Planetology

Examination of Mars's craters reveals in part the history of modification of Mars's surface. Images transmitted from the first spacecraft sent to fly by Mars revealed a cratered surface somewhat resembling the lunar surface rather than one that might have been more Earth-like. However, craters on Mars undergo weathering processes that are different from those on either Earth or the Moon.

Overview

There are two classes of craters on any planet or other terrestrial body: impact craters, caused by meteors or comets crashing into the surface, and volcanic craters. On all the known bodies of the solar system, including Mars, impact craters outnumber volcanic craters by the thousands.

Early views of the planet Mercury and the detailed knowledge of Earth's moon depicted both bodies as heavily cratered. Most of this cratering occurred in the earliest 500 million years, after the planets had formed by accretion out of the primordial material that constituted the early solar nebula and Sun. During this period, the planets were heavily bombarded from space by meteors and planetesimals caught in the gravitational pull of the newly forming planets.

Unlike Mars, Mercury, and its own moon, Earth does not show evidence of having been heavily cratered, although it was subjected to the same rate of incoming meteors; Earth's thick atmosphere burns up and destroys any incoming body of less than 1,000 kilograms in mass. Earth's widespread weathering processes also quickly erase evidence of meteoritic craters. Any trace of a meteor falling into the ocean is either totally erased or largely mediated by the water. On airless, inactive bodies, however, there are few mechanisms to erase cratering, even though some occurred billions of years ago.

The biggest craters on Mars, typically caused by very large impacting bodies (such as asteroids), are called basins. The largest such basin on Mars is called Hellas Planitia; with a diameter of 1,600 kilometers and a depth of 3 kilometers, it is the deepest point on the planet's surface. (As large as it is, it is still not the largest known in the solar system. The lunar basin called Oceanus Procellarum—the Ocean of Storms—has a diameter of some 2,500 kilometers.) Other such large features are Isidis (1,400 kilometers) and Argyre (900 kilometers).

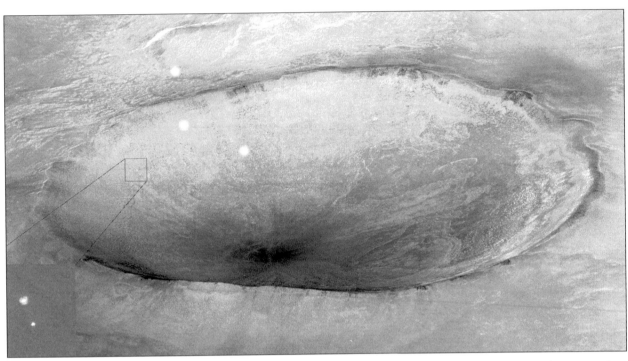

The High Resolution Imaging Science Experiment (HiRISE) relayed this image of the Mars Phoenix Lander, attached to its parachute (enlarged view lower left corner), as it drifted down to the landing site in Mars's arctic region on May 25, 2008. The 10-kilometer-diameter Heimdall Crater occupies most of the image area. (NASA/JPL-Caltech/University of Arizona)

The most common type of crater found on Mars is called a rampart crater, first described by planetary geologist J. F. McCauley in 1973. Rampart cratering seems almost unique to Mars. This interesting morphology consists of a central crater wall with an ejecta pattern that resembles a resolidified flow pattern radiating outward from the crater wall as a low ridge and radial striae. This flow pattern hints that the impact actually liquefied the subsurface materials, causing them to flow away from the impact point, then resolidify as a rampart, or gently sloping wall. Causes of this fluidized ejecta are thought to have been entrapment of atmospheric gases in the ejecta or, more significantly, water in the ejecta material, such as permafrost, which acts as a lubricant, allowing the materials to flow away from the impact point. The latter concept is a favorite one of those who hope to discover immense amounts of water locked up in subsurface deposits as permafrost. Since rampart cratering is prevalent on much of Mars, such planetwide deposits of permafrost would be a very positive sign for eventual human exploration of the planet.

One of the most unusual craters found on Mars is the pedestal crater, formed by weathering action that cuts away at the base of the crater until the crater takes on the appearance of a pedestal. The unique meteorology and geology required to produce such craters are found on Mars between 40° and 60° north latitude.

In examining craters on Mars, planetary scientists are able to define the crater's age, and hence that of the surrounding terrain, through a process of observing certain of its characteristics. The process of a crater eroding away is called degradation. Examining the original impact that formed the crater is a starting point; a new crater with little or no degradation is one with sharp edges and fresh ejecta outlying its central diameter. As a crater degrades through natural processes, its rim loses sharpness, and crater walls slump and lose their definition. Ultimately, the walls may form gullies and eventually become completely degraded to surface level. In the final stage, the crater is hardly noticeable over the terrain, and some planetary scientists even call such a crater a ghost.

On Earth, the most profound degradation process is weathering, which can erase even large craters relatively quickly, in geologic terms. On Mars, such processes occur much more slowly. Since liquid water does not presently flow on the Martian surface, weathering processes are confined to wind erosion, or aeolian (wind) processes. It is estimated that aeolian erosion is responsible for filling in craters on Mars at the rate of 0.0001 centimeter per year. From these crater studies, there is considerable evidence that such erosion has slowed in the planet's recent geologic past. The last period of relatively heavier erosion occurred roughly 600 million years ago, according to some crater studies.

The interior morphology of Martian craters is highly variable. It is typically dependent on the mass of the incoming projectile and the composition of the impact site. In smaller craters, the bottom usually assumes a more spherical shape. In larger craters, the central region of the crater flattens out until, in craters greater than 25 kilometers in diameter, a central peak is formed. Such flatness is often caused by the impacting body liquefying the crust or by magma welling up from the planet's interior to fill in the crater.

By examining craters formed within other craters in a process called superimposition, planetary geologists are able to determine the age of whole regions of Mars's surface. With an overall planetary comparison, which includes mass crater counts, it has been discovered that there are areas that have changed little since Mars was formed, while other regions, notably volcanic regions, have changed in the recent geologic past.

In some of the most fascinating crater studies accomplished to date, scientists have attempted to determine the age of what appears to be massive river channels on Mars. Although conditions on Mars at present will not allow liquid water to exist on the planet's surface, it appears that water once flowed in an extensive series of channels. Planetary geologists have examined craters that overlie the enigmatic channels. They range in age from 3.5 billion years to as young as fewer than 200 million years, strongly suggesting that the channels have been formed on a cyclic basis throughout the geologic history of Mars.

Mars has the largest volcanic craters in the known solar system, resulting from volcanoes between ten and one hundred times larger than the largest volcano on Earth, Mauna Loa in Hawaii. Four very large volcanic craters exist in close proximity in the Tharsis region. The largest volcano is Olympus Mons, a 200-million-year-old volcano with a crater 80 kilometers across. These Martian volcanoes, called shield volcanoes after their earthly counterparts, have craters nested inside one another. Three other Tharsis volcanoes are roughly alike in size, all much larger than Mauna Loa. The crater on Arsia Mons is 140 kilometers in diameter (compared to

Mauna Loa's 2.8 kilometers). The largest Pavonis Mons crater is 45 kilometers in diameter and 5 kilometers deep (compared to Mauna Loa's 0.2 kilometer of depth). The largest of the Ascraeus Mons craters is roughly the same diameter as the largest Pavonis Mons crater. The volcanic plains spreading away from the Martian volcanoes are dated according to crater distribution. The youngest is Pavonis Mons, at 80 million years.

Methods of Study

Earth-based telescopic observations of Mars cannot clearly reveal its distinct surface features. The distance is too great and the combined effects of Earth's and Mars's atmospheres reveal only indistinct and blurred splotches of color. The means for collecting aggregate crater data is basically orbital photography from spacecraft. Almost all such data have come from flyby spacecraft, orbiters, and landers.

The first distinct images of the planet's surface came from NASA's Mariner 4 spacecraft on July 14, 1965. This robotic spacecraft transmitted nineteen photographs of the Martian surface back to Earth. In the twenty-two minutes that it took to transmit the photographs, previous speculation about the Martian surface was laid to rest. The most surprising and pronounced of the Mariner 4 revelations was the extensive cratering of the Martian surface, a finding few planetologists had expected. Popular interest in Mars waned at this point, however, because the planet appeared to be rather Moon-like instead of perhaps a place where life might have had a chance to develop.

Based on observations from Earth-based telescopes, it was widely expected before 1965 that there was enough weathering on Mars to have erased most of its primordial cratering. Mariner 4 clearly showed that this was not the case. The most immediate implication was that Mars was far less active meteorologically and geologically than was initially thought. Later, both extreme views were mediated by data from spacecraft that gave rise to more detailed studies, which showed that Mars has an active weathering process, though not as vigorous as that of Earth.

The twin Mariner 6 and 7 spacecraft flew past the Red Planet only a few weeks after the Apollo 11 astronauts returned to Earth following the first crewed landing on the Moon. At this point, interest in sending humans to Mars was at a precarious stage: Some were thrilled that Apollo had achieved its goal but now were no longer interested in pursuing expanded human-based scientific exploration of space; others were eager to build on the success of Apollo and fly to Mars during the 1980's. The supposition

Victoria Crater, about 800 meters in diameter, near the Martian equator, as imaged by the Mars Exploration Rover on May 6, 2007. (NASA/JPL/Cornell University)

of many was that the picture of Mars as revealed by Mariner 4 was atypical of the entire planet. However, both Mariners 6 and 7 sent back a catalog of images of the Red Planet that dashed the hopes of those expecting Mars to be more Earth-like. The majority of the images displayed cratered features with no evidence of surface water.

NASA program designers in the 1960's included sending a pair of spacecraft toward a planetary target to ensure mission success if a launch or other accident resulted in the loss of one. The wisdom of that policy had been demonstrated when Mariner 1 suffered a launch accident but Mariner 2 encountered Venus. In early Mars exploration, Mariner 3 had failed but Mariner 4 was a huge success. Both Mariners 6 and 7 were successful, but of the pair of orbiters in the Mariner Mars next sent to Mars by NASA, Mariner 8 suffered an upper stage shutdown 265 seconds after liftoff and fell back into the Atlantic Ocean. Mariner 9, however, became the first spacecraft to enter orbit about another planet on November 13, 1971.

Whereas Mariners 4, 6, and 7 had revealed an inactive world more like the Moon, pockmarked with craters,

Mariner 9 returned images with resolutions as good as 0.1 kilometer per pixel and thus revealed an entirely new side of Mars's surface characteristics. In addition to craters, Mariner 9 provided evidence of large-scale flows of water across the planet's surface in the distant past. Volcanic craters were imaged to greater resolution, but no evidence was obtained to indicate present-day volcanic activity or hot spots near Mars's huge volcanoes. The images and data from Mariner 9 helped scientists determine how and when to attempt the next stage of Mars exploration: sending landers to the surface in 1976.

Viking 1 and 2 each consisted of a joint orbiter and lander. Whereas the orbiters continued to image as much of the Martian surface as possible—including craters, valleys, volcanoes, and polar regions—the primary focus of the landers was to search for conditions that might support in the present, or might have supported in the past, primitive life on Mars. Viking 1 landed on Mars on July 20, 1976, in western Chryse Planitia, and Viking 2 on September 3, approximately 200 kilometers to the west of the crater Mie in Utopia Planitia. Both landers analyzed the soil but failed to find any organic material or evidence of life. It would be twenty-two years before another successful powered landing of a spacecraft on the Martian soil. Mars Phoenix successfully touched down on the Red Planet's northern arctic region on May 25, 2008, in an area largely devoid of large rocks. The landing site was chosen because it offered a high possibility of the spacecraft's encountering permafrost with a layer of ice either on the surface or not very deep beneath.

Russian plans for extensive exploration of Mars ended with the Phobos 1 and 2 spacecraft in 1988 and 1989. One failed en route to Mars, the other as it approached the satellite Phobos. With the dissolution of the Soviet Union in 1991, Russian plans for Mars programs dissolved as well.

Between the Viking (1976) and Mars Phoenix (2008) landings, NASA refined its plans for Mars exploration at least twice. As a result of the indication that water had once flowed across the cratered Martian surface, a new wave of Mars spacecraft were sent to the Red Planet. It began with Mars Observer, which unfortunately exploded as it prepared to execute its Mars orbit insertion burn. Mars Observer was followed by Mars Pathfinder, a surface rover; Mars Global Surveyor, an orbiter; Mars Climate Orbiter, which failed to achieve orbit and burned up in Mars's thin atmosphere; Mars Polar Lander, which crashed in the southern polar region; Mars Odyssey, an orbiter; twin Mars Exploration Rovers named Spirit and Opportunity; the European Space Agency's Mars Express; the Mars Reconnaissance Orbiter; and the Mars Phoenix, a powered surface lander. The thrust of scientific investigation of all of these second-generation Mars probes was a search for Martian water, but that required investigation of Martian geology. This search included the possibility of water ice deposits in craters and the erosional aspect of water on crater walls.

In the process, these spacecraft provided greatly improved images of Mars. Some returned data over long periods of time, including detailed meteorological observations from both orbit and the surface. From this information, Martian cratering and morphology were assessed in detail. Some of the most elemental assessments of crater morphology include crater size, composition of ejecta, composition and behavior of impacted terrain, modification processes of the crater, and its effect on surrounding craters and terrain. These findings led to a determination of the ages of the craters, surrounding crater fields, and even the impacted terrain. Assessments were

made of crater sizes, numbers, distribution, and ages within selected planetary areas.

Techniques for analysis of Martian craters vary with the aim of the study. For example, a statistical counting method borrowed from other applications, called a frequency distribution, is used to determine the number of craters located over a given area. This information may lead to determination of the age of the area under study, given a uniform crater deposition rate. Other statistical counting methods include cumulative distribution, logarithmic incremental distributions, and incremental frequency distributions, all of which are specific methods of presenting collected crater population data. Some appear as graphs, others as numerical tables, and others as simple maps showing locations of craters over a given area. In the largest distribution sampling exhibit, the entire planet of Mars is presented with its crater distributions marked. All of these are unique representations designed to give the researcher information in a specific way so that the concept under study, such as the age of an area, can be efficiently gauged. Inferences are made from these statistical presentations about such broad concepts as crater production and erosion.

Planetologists have arrived at a rate of cratering for Mars that is equivalent to three hundred meteors, of one kilogram or larger, striking each square kilometer every million years. From this baseline of impacts, current photographs can be compared to establish which areas are very old and which are younger geological formations. The logical extension of such knowledge is the ability to date such formations as plains, volcanic flows, and streambeds, as well as to establish a baseline for regional aeolian erosion.

Context

The study of Martian craters allows planetary scientists to determine from orbital photographs many varied characteristics of Mars. From information relating present conditions on Mars all the way back to its earliest geological history, crater studies have allowed much of the planet's chronology to be traced. Orbital studies of Mars's craters may lead to answers about the planet's most significant geological history, such as whether water ever existed on Mars in liquid form, what formed the vast river channels and canyons, and what happened to the water. Answers to such questions will determine what humanity will have to do to survive on Mars should this neighbor planet be visited and perhaps colonized, as proposed in the Bush administration's 2004 Vision for Space Exploration.

Crater studies also address questions of importance relative to Earth's own history and future. Planetologists seek to understand why Mars is so different from Earth and what indeed happened to its once apparently abundant water. Such questions of planetary history relate to what may one day happen on Earth. If the scientific community is eventually able to draw enough parallels from the study of other planets, it may be able to apply rigid mathematical projections of Earth's own future based on present conditions and trends.

Dennis Chamberland

Further Reading

American Geophysical Union. *Scientific Results of the Viking Project.* Washington, D.C.: Author, 1978. A compendium of articles about the Viking project, originally published in the *Journal of Geophysical Research.* It consists of detailed assessments of data from the Viking landers and orbiters. Suitable for college-level readers.

Barlow, Nadine. *Mars: An Introduction to Its Interior, Surface, and Atmosphere.* Cambridge, England: Cambridge University Press, 2008. An interdisciplinary text including contemporary data from the Mars Exploration Rovers and Mars Express. Each chapter contains necessary background information. A good reference for planetary science students.

Beatty, J. Kelly, Carolyn Collins Petersen, and Andrew Chaikin, eds. *The New Solar System.* 4th ed. Cambridge, Mass.: Sky, 1999. Filled with color diagrams and photographs, this popular work on solar-system astronomy covers planetary exploration through the Mars Pathfinder and Galileo missions. Accessible to the astronomy enthusiast.

Collins, Michael. *Mission to Mars.* New York: Grove Weidenfeld, 1990. Apollo 11 astronaut Collins provides an astronaut's vision of a trip to Mars. Examines the problems to be overcome to make such a journey possible with space-shuttle-era technology.

De Pater, Imke, and Jack J. Lissauer. *Planetary Sciences.* New York: Cambridge University Press, 2001. A challenging and thorough text for students of planetary geology. Covers extrasolar planets as well as an in-depth explanation of solar-system formation and evolution. An excellent reference for the most serious reader with a strong science background.

Ezell, Edward, and Linda Ezell. *On Mars: Explorations of the Red Planet, 1958-1978.* NASA SP-4212. Washington, D.C.: Government Printing Office, 1984. The classic official NASA history of the Viking program from the original ideas in 1958 to the culmination of the project some twenty years later. Provides a detailed assessment of the political history and discusses details of the instruments that photographed Mars from orbit as well as the surface instruments that measured the winds. Generally nontechnical and accessible to most readers.

Harland, David M. *Water and the Search for Life on Mars.* New York: Springer Praxis, 2005. A historical review of telescopic and spacecraft observations of the Red Planet up through the Spirit and Opportunity rovers. Covers all aspects of Mars exploration, with emphasis on the search for water.

Hartmann, William K. *Moons and Planets.* 5th ed. Belmont, Calif.: Thomson Brooks/Cole, 2005. An updated version of a classic text on planetary science. The material on Mars covers all aspects of ground-based and spacecraft observations of Mars. Takes a comparative planetology approach instead of including separate chapters on each major solar-system object.

Squyres, Steve. *Roving Mars: Spirit, Opportunity, and the Exploration of the Red Planet.* New York: Hyperion, 2006. Written by the principal investigator for the Mars Exploration Rovers Spirit and Opportunity, this fascinating book provides a general audience with a behind-the-scenes look at how robotic missions to the planets are planned, funded, developed, and flown. A personal story of excitement, frustrations, a scientist's life during a mission, the satisfaction of overcoming difficulties, and the ongoing thrills of discovery.

Zubrin, Robert. *Entering Space: Creating a Spacefaring Civilization.* New York: Tarcher, 2000. The author displays a gung-ho attitude toward making humanity a truly spacefaring species by accepting the challenge of journeying to Mars sooner rather than later using contemporary technology and daring innovation. Speculates beyond travel to Mars.

MARS'S POLAR CAPS

Categories: Mars; Planets and Planetology

In 1666, Gian Domenico Cassini observed Mars's surface and described two polar caps, the planet's most visible features. These caps have been studied extensively since, both telescopically and by means of the Mariner, Viking, Mars Global Surveyor, Mars Odyssey, Mars Reconnaissance Orbiter, and Mars Express space probes.

Overview

When Gian Domenico Cassini observed the polar caps of Mars in 1666, little was known about the planet's surface features. A decade before Cassini, the Dutch astronomer Christiaan Huygens suggested the presence of polar caps on Mars, but it was not until 1672 that he actually saw the planet's south polar cap, Mars's most apparent feature when one views it through a telescope. No one before Cassini had seen the polar caps in detail, because no instrument had existed to allow sufficient magnification.

After the invention of the telescope, early astronomers who observed some of the surface features of Mars reached conclusions based on analogies with Earth, the only planet whose atmosphere and chemical composition they knew in detail. It is not surprising, then, that the Martian polar caps were presumed to be like the polar caps found on Earth: composed of ice, snow, hoarfrost, or a combination of these forms of water. This theory presupposes that Mars has water, considered a requisite for life, and therefore the individual components of water, hydrogen and oxygen. This notion gave rise to the popular theory that Mars could support life in some form, a theory that was subsequently brought into considerable question and that remained controversial after the early Mariner and Viking missions to Mars.

In 1719, the Italian astronomer Giacomo Maraldi discovered that the polar caps are not centered at Mars's precise geographical poles. Maraldi also divined from his observations that the south polar region of Mars has a much greater mass than its northern

counterpart. Subsequent research and observation have substantiated Maraldi's theories, revealing that the planet's north polar cap lies about 64 kilometers from its geographical north pole and that the south polar cap lies some 400 kilometers from the geographical south pole. This phenomenon occurs because of the planet's elongated orbit.

Italian astronomer Giovanni Schiaparelli trained telescopes on Mars repeatedly during the 1877 opposition of the Red Planet and drew highly detailed maps of Mars's surface. Those maps documented networks of linear features Schiaparelli was convinced existed in great number between 60° north and south in latitude. Reports of these linear features were taken out of context in public accounts. Schiaparelli had termed these features *canali*, which from Italian translates as "channels." However, these linear features were most often referred to as "canals," with the obvious implication that they were artificial structures (in turn implying that they were built by intelligent life-forms) to convey water from one location on the planet to another. Thus the notion of canals on Mars was born and proceeded to take on a life of its own in science fiction as well as in pseudoscience.

Olympus Mons, the tallest volcano on Mars, appears as a bump in the lower center of this image from the Mars Global Surveyor, while the north polar region displays an icy crown. (NASA/JPL-Caltech/University of Arizona)

The wealthy astronomer Percival Lowell developed a fascination with the planet Mars. He built the Lowell Observatory in Flagstaff, Arizona, atop Mars Hill and spent the final twenty-three years of his life deeply engaged in astronomical investigations, especially focusing on Mars. Lowell sought to use advanced telescope technologies to search for the canals of Mars prominent in Schiaparelli's maps of the Red Planet. Lowell believed in and popularized with the public the notion that the canals were evidence of alien technology. In three books, Lowell advanced the theory that the Martians had built an elaborate system of canals to bring precious water from the polar ice caps to population centers suffering from increasingly arid conditions.

Lowell's notions were not well received by the scientific community (although he later would expend considerable effort to locate a body beyond Neptune, which bore fruit after his death with the discovery of Pluto by Clyde Tombaugh). The present atmosphere of Mars is known to be such that there can be no accumulated areas of free liquid water, such as lakes, ponds, rivers, or oceans on the planet's surface. Much evidence suggests, however, that in the past the Martian surface was extensively cut by flowing liquid. Most evidence demolishes the theory that this liquid was lava flowing from the planet's once active volcanoes. Current theory indicates, rather, that in one stage of the planet's evolution, water in liquid form was plentiful.

When everything astronomers knew about Mars came from their observations through telescopes, they could gather substantial information about the polar caps; they were at odds, however, in their interpretations of these data. They could not always be sure what they were seeing. They also had little means of knowing with any certainty the depth of the polar caps. Some scientists thought that they were hundreds of feet deep; others thought the caps were merely thin coverings of hoarfrost. Common sense supported the latter idea. The argument was that if a thin layer of hoarfrost covered the poles, it would, under warmer conditions, condense and form clouds. On the other hand, if the polar caps were composed of fairly deep layers of solid water, where would the melting residue run during the warmer season? The atmosphere of Mars was then thought to be almost totally dry, yet observations through telescopes clearly showed that in the summer, the polar caps seemed to melt partially. Certainly they diminished in size, and the areas surrounding them assumed a dark coloration, suggesting that water was mixing with minerals and dust at the periphery of the thawing area and causing what looked like a moist condition.

The south polar cap covers an area of more than 10 million square kilometers in winter and at times extends almost halfway to the Martian equator. Even a minimal thawing would produce water in great quantity, especially if the water ice ran to great depths; yet, no water from the polar cap appeared through the telescope to have penetrated other areas of the planet, which seemed dusty and dry.

Faced with this anomaly, scientists could do little to verify their theories until a different sort of evidence, the

This panorama of a plain in Mars's north polar region shows a 360° view around the Mars Phoenix Lander and surface patterns reminiscent of those seen in permafrost areas on Earth. (NASA/JPL-Caltech/University of Arizona/Texas A&M University)

kind first provided by the Mariner and Viking missions, became available in quantity between 1965 and 1977. Viking 2 measured some of the temperatures of the north polar cap with a high degree of certainty and discovered that temperatures in its dark surrounding areas had a range of between 235 and 240 kelvins, and that the white, presumably frozen areas had a range between 210 and 205 kelvins. These temperatures exceed 194 kelvins, the point at which carbon dioxide sublimates into a solid.

The atmosphere of Mars, thought until the Mariner 4 flyby in 1965 to be about 85 millibars, was proved to be a thin 10 millibars; Earth's atmosphere at sea level is 1,000 millibars, or 1 bar. The Martian atmosphere is comparable to Earth's atmosphere at about 24,000 meters, where the pull of gravity is all but lost. Such an atmosphere does not permit freestanding water, indicating that if the Martian atmosphere had always been the way it is now, the planet could never have had water, and the polar caps would necessarily be composed of some other substance.

This sort of thinking gave rise to the theory that the polar caps were composed of solid (frozen) carbon dioxide, or dry ice. Lowell had advanced such a theory as early as 1895; other eminent astronomers considered it more likely than the water theory as late as 1971, when Mariner 9 was uncovering data that would soon vitiate, although not completely eliminate, the carbon dioxide theory. Mariner 4, when it made its flyby of Mars in 1965, had returned data suggesting that the Martian atmosphere was composed largely of carbon dioxide under weak pressure. This information caused some astronomers to cast their lots with Lowell's theory that carbon dioxide in its frozen state covered the polar caps but that, when the temperature rose, it became, through sublimation, gaseous and returned to the atmosphere as mist or fog. This explanation helped to account for the haze often observed over the polar regions.

Mariner 9 data substantiated the theory that Mars's present atmosphere accommodates no accumulations of liquid water. It also presented incontrovertible evidence that the polar caps are composed largely of ice and are minimally 0.8 kilometer deep, indicating

clearly that the Martian climate has changed through the eons and that it was once such that liquid water, now locked in the polar caps in solid form, was abundant.

The earliest space missions carried out research only concerning the north polar cap, but they managed to dispel a substantial number of misconceptions about Mars, among them the mistaken idea that the planet is totally dry. Data sent back suggested that in the northern latitudes above 60°, the atmosphere is quite moist; the atmosphere over the north polar cap has twenty times the water vapor that the atmosphere over the equatorial regions contains. During the Martian summer, surface ice exposed to the Sun evaporates in the morning, causing a mist that seems to condense, resulting later in the day in precipitation. Although these data were gathered exclusively over the north polar cap, there was little reason to think that the south polar cap differed significantly. At perihelion, the planet's closest approach to the Sun, the north polar cap does not face the Sun directly but is tilted away from it. This tilt prevents it from getting the full impact of the Sun's rays, which would, presumably, melt a polar cap composed of frozen water.

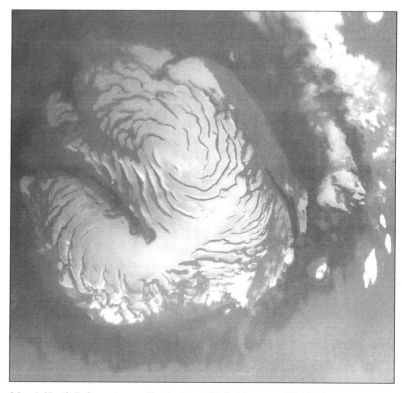

Mars's North Pole, as imaged by the Mars Global Surveyor. (NASA)

Space probes to Mars indicated that in the ancient past, water must have been quite plentiful on the planet. The climate of Mars must be viewed over eons; the planet is now much drier than it once was, and scientists think that it could evolve through this period to one in which conditions resemble what they were when the Martian atmosphere allowed the accumulation of water in bodies similar to the ones on Earth.

It is now widely accepted that portions of both of the major early theories about the polar caps were true. Although the evidence is strong that deep layers of water ice cover the polar regions and although it is known that the deep crater Korolev in the north polar cap is filled with water ice, many astronomers think that during certain seasons there are thin coatings of carbon dioxide that sublimate into the gaseous state and cause the clouds or mists that have been observed over the polar caps.

Methods of Study

In the three and a half centuries that the polar caps of Mars have been observed, descriptions of the caps have moved from highly speculative to soundly scientific. In this period, astronomers have moved from primitive optical telescopes to incredibly complex telescopes of enormous size and capable of detecting invisible radiation, strategically placed to focus on the sky and on the planets. Mars has been the most intriguing planet for most astronomers to explore, because it is sometimes a relatively close 56 million kilometers from Earth. It is also the planet that most resembles Earth in its surface features, although its atmosphere currently precludes advanced life.

The Mariner and Viking missions first solidified human knowledge of Mars. Telescopic evidence suggested to many that, because of reflections detected from the polar caps, the caps must be composed of water ice. When the earlier Mariner missions presented evidence favoring the carbon dioxide theory, astronomers were forced to rethink their earlier stands. Later expeditions, however, offered convincing evidence in favor of the water ice theory. When Viking 1 landed, it not only photographed the surface extensively and transmitted the pictures to Earth but also

deployed an arm that dug into the Martian surface and analyzed the composition of the materials it uncovered. Finding water locked in Martian rocks established clearly the former existence of water on the planet.

Mariner 9 was placed into orbit around Mars in 1971, and it sent back more than seven thousand pictures. It photographed the south polar cap continuously from November, 1971, until March, 1972, capturing the waning of the polar cap as summer advanced. These pictures provided extremely varied information and reiterated the water ice theory.

The goal of the Viking program was to land on the surface of the planet and analyze the soil for signs of life. However, it is often overlooked that the landers were each transported to Mars by an orbiter. The Viking 1 orbiter lasted from orbital insertion on June 19, 1976, until August 17, 1980, when it ran out of attitude control propellant. The Viking 2 orbiter lasted from August 7, 1976, until July 25, 1978, at which point it suffered a fuel leak. During those periods, these orbiters produced extensive catalogs of images of a large majority of the Martian

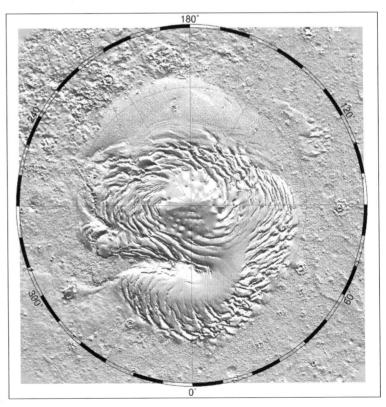

This model displays the topography of Mars's north polar region and ice cap. (NASA/JPL/GSFC)

surface, including photographs of both polar regions, which over time displayed seasonal changes.

The Mars Observer spacecraft was launched on September 25, 1992, and was designed, among other things, to generate a global distribution map of Mars's elements and minerals. This spacecraft was one of the largest and most sophisticated (for its time) to be dispatched to Mars. Unfortunately, before it could enter orbit and perform any scientific operations, including studies of the polar caps, Mars Observer was lost on August 21, 1993. However, its loss resulted in a new approach to interplanetary exploration, National Aeronautics and Space Administration (NASA) Administrator Daniel S. Goldin's infamous "faster, better, cheaper" policy, which sought to replace expensive, large-scale probes such as the Mars Observer with more focused, less expensive probes sent out into the solar system with greater frequency and more risk by using cutting-edge technologies. "Faster, better, cheaper" turned out, as evidenced by the subsequent losses of Mars Climate Orbiter and Mars Polar Lander and others to be neither better, faster, nor even cheaper in the long run. Goldin's approach managed to delay by almost a decade what had been a

well-thought-out program of Mars exploration intended to robotically return rock and soil samples to Earth by the middle of the first decade of the twenty-first century. The impact of Mars Observer also had serious repercussions for studies of the Martian polar regions.

Mars Global Surveyor (MGS) was launched on November 7, 1996, and entered a stable Martian orbit on September 11, 1997. It was the first successful American spacecraft to the Red Planet in twenty years since the two Vikings. MGS entered a near-polar orbit, so that as it imaged geological features across the planet it also was used to study the polar ice caps. Its purpose, to search for evidence of water, went far beyond merely searching for water in the polar regions. MGS provided images of gullies on the walls of craters that could be the result of groundwater eroding the crater walls before freezing. In early November, 2006, MGS went silent after experiencing a problem in directing its solar arrays toward the Sun to generate sufficient electrical power.

The Mars Climate Orbiter (MCO) was NASA's next probe sent to the Red Planet after the highly successful Mars Pathfinder rover, which generated tremendous

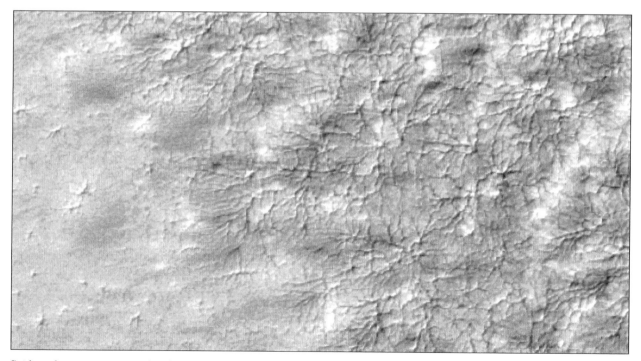

Spiderweb patterns emerge after the seasonal disappearance of the Martian north pole's carbon-dioxide ice cap. (NASA/JPL/Malin Space Science Systems)

national and international excitement in 1997. MCO was launched on December 11, 1998, to study Mars's atmosphere primarily, but also, by investigating the transport of water and atmospheric circulation, it would have indirectly increased humanity's understanding of the polar caps. Unfortunately, the result of a human miscommunication between engineers, some working using the metric system and others using British units, led to MCO's coming in too close to Mars on final approach to orbital insertion. On September, 23, 1999, MCO performed its Mars orbit insertion burn but never came around the back side of the Red Planet. Entering orbit at far too low an altitude, only 57 kilometers above the surface, MCO burned up and was lost.

The Mars Polar Lander (MPL) was a powered lander intended to touch down on the surface of the Red Planet in the southern polar region in order to search for evidence of water in the soil. The spacecraft was equipped with a sample scoop at the end of a robotic arm to dig up samples. A gas analyzer was included among its scientific instruments in order to heat soil samples and detect any volatile gases, such as water vapor, given off in the process. On December 3, 1999, MPL began its entry phase at a speed of 68 kilometers per second. Unfortunately, the onboard guidance system appeared to cut off the engines, slowing its speed at an altitude of 40 meters and causing the lander to crash on the Martian surface.

Mars Odyssey was launched on April 7, 2001, and entered a nearly polar orbit on October 24, 2001. In addition to being a scientific research platform, the spacecraft was intended to serve as a communications relay for later surface probes (such as Mars Phoenix). Designed primarily to search for evidence of water, perhaps the most important discovery made by Mars Odyssey was detection of enormous amounts of hydrogen, which was interpreted as an indication of subsurface water. An extended mission was approved to continue observations of the polar ice caps for contemporary changes.

Mars Express was launched by the European Space Agency (ESA) on June 2, 2003. It represented the first planetary mission attempted by ESA and was followed a few years later by a similarly designed probe sent into Venus orbit, the Venus Express. Mars Express entered Mars orbit on December 25, 2003. It carried a radar system capable of detecting subsurface water ice in permafrost regions. In a near-polar orbit, it was able to study surface changes and make mineralogical maps in addition to monitoring seasonal changes in the polar caps. Mars Express's Fourier spectrometer discovered that methane was entering the atmosphere in near-equatorial regions having subsurface ice. Its Visible and Infrared Mineralogical Mapping Spectrometer detailed hydrated sulfates. Radar data provided evidence of underground water ice. Mars Express found the signature of water ice in the planet's south pole. What some interpreted as a deposit of water ice was discovered on the floor of an otherwise nondescript crater in the vicinity of Mars's north pole.

Mars Reconnaissance Orbiter (MRO) was launched on August 12, 2005, and injected into orbit about Mars on March 10, 2006, joining five other operational spacecraft already performing scientific investigations of the Red Planet. MRO was outfitted with the largest imaging system (HiRISE) ever sent to another planet, as well as spectrometers and a penetrating radar. Among its primary science objectives were searches for water in the

The Martian south pole's characteristic "swiss cheese" terrain, carved into the carbon-dioxide ice cap. (NASA/JPL/Malin Space Science Systems)

polar regions and a daily determination of meteorological conditions that might lead to changes in the state of Mars's water deposits. HiRISE images provided evidence of banded terrain, suggesting the action of water on the surface in relatively recent geological times. The spacecraft's radar was able to detect underground deposits of water ice.

Mars Phoenix successfully touched down on Mars's northern arctic region on May 25, 2008, in an area largely devoid of large and medium-sized rocks. Its landing marked the first time since the Viking spacecraft that a powered lander had successfully reached the Martian surface. The landing site was chosen to have a high possibility of a layer of ice either on the surface or not very deep beneath. Mars Phoenix was outfitted with a robotic arm, at the end of which a scoop was attached. The scoop not only could pick up loose surface soils but also could scrape the ice. The idea was to lift soil and potential ice into a special gas analyzer that would process the soil and chemically analyze it for the presence of water and organic molecules. Very soon after Mars Phoenix landed, initial photographs strongly hinted that the lander had touched down near ice. Under the lander it appeared that exhaust from Mars Phoenix's rockets had kicked up the soil, exposing subsurface ice. Early attempts to analyze the soil were thwarted by communications problems, and the entryway into the gas analyzer was clogged by soil that clumped. The very first run of the lander's Thermal Evolution and Gas Analyzer produced a disappointing lack of any water signature. However, subsequent data did provide strong evidence of water at the lander's site.

Also in May, 2008, scientists studying MRO data revealed that it appeared that Mars's crust and upper mantle could be even colder than originally believed. If true, that would mean that liquid water would have to be found even deeper, where internal heat would permit water to exist in that phase. MRO's Shallow Radar experiment probed the internal structure of ice, sand, and dust layers on the north polar cap and revealed the planet's lithosphere (combination of crust and mantle) to be thicker and colder than previous models indicated, and as such provided greater support against the stress of the weight of the polar cap atop it. Radar imagery indicated four zones of thin ice and dust layers interspersed with rather thick layers of water ice going down into the lithosphere to a geological depth such that it recorded changes in climate on Mars over many millions of years. The thin layers represented climate changes perhaps lasting only one million years. These data did not shed light on the cause of the climate change, but two strong possibilities were alterations of Mars's rotational axis and/or orbital eccentricity.

Context

The question whether there is now or could ever be life on Mars has long been a matter of conjecture. If there is, was, or will be life on Mars, that life would likely be confined to organisms much smaller and less complex than anything resembling human beings. To sustain animal or vegetable life, it is presumed, some minimal atmosphere must exist and water must be present. Research into the polar caps of Mars provides evidence that water is plentiful in the planet's frozen polar caps. The same research suggests that the planet is much drier than Earth, that its atmosphere is so rarefied that it cannot support complex organisms, and that its temperatures (although much less forbidding than those on Jupiter or Venus) are not conducive to life.

In the second quarter of the twentieth century, Eugene M. Antoniadi, a Greek-born astronomer who spent his professional life in France, explained the dark areas on the periphery of the polar caps in summer, designated the "Lowell bands" for astronomer Percival Lowell. Antoniadi thought that the fringes of the ice fields were reduced in brightness by grasses and bushes that grew in the periphery, presupposing by such a contention the existence of life on Mars and of an atmosphere that would support life. This theory has been disproved, but it reflected the widespread notion that life exists on Mars—although not in the form of the "little green men" that some works of science fiction describe.

Spacecraft exploration of the polar caps of Mars has revealed that water in some form exists on Mars and has suggested that water in liquid form was once more plentiful there than it is currently. The channels of Mars are now generally thought to have been forged through the eons by flowing water, and the presence of the polar ice caps supports this theory. These explorations have also presented evidence that the atmosphere of Mars has changed drastically from what it once was. Climatic change may one day produce a Mars quite different from the present planet. Astronomers who make such projections, however, caution that they are talking about millions, perhaps hundreds of millions, of years. Mars is unlikely to change in easily perceptible ways within the life span of a single human now living.

In the aftermath of the 2003 space shuttle *Columbia* accident, President George W. Bush redirected NASA to complete the International Space Station and retire the shuttle fleet by 2010 in order to move on to human exploration beyond low-Earth orbit. Project Constellation was intended to take astronauts back to the Moon by 2020 to begin a permanent research outpost there and in time to send human explorers to Mars. In order to make both of those goals happen, the use of local water would be required. Sending humans to Mars would be possible only if the robotic search for water on Mars bore fruit; the polar ice caps would play prominent roles in liberating water from the surface of Mars for use by future astronauts.

R. Baird Shuman

Further Reading

Barlow, Nadine. *Mars: An Introduction to Its Interior, Surface, and Atmosphere*. Cambridge, England: Cambridge University Press, 2008. An interdisciplinary text including contemporary data from the Mars Exploration Rovers and Mars Express. A good reference for planetary science students and nonspecialists alike. Each chapter contains necessary background information.

Carr, Michael H. *The Surface of Mars*. Cambridge, England: Cambridge University Press, 2007. The author provides a complete description of the geological heating of Mars as understood based on results from Mars Global Surveyor, Mars Odyssey, Mars Reconnaissance Orbiter, Mars Express, and the Mars Pathfinder and Mars Exploration Rovers. Heavily illustrated; includes a comprehensive reference list.

Encrenaz, Thérèse, et al. *The Solar System*. New York: Springer, 2004. A thorough exploration of the solar system from early telescopic observations through the space missions that have investigated all planets with the exception of Pluto by the publication date. Takes an astrophysical approach to give our solar system a wider context as just one member of similar systems throughout the universe.

Glasstone, Samuel. *The Book of Mars*. Washington, D.C.: National Aeronautics and Space Administration, 1968. This classic work contains a useful history of Mars up through the early Mariner flyby missions. Provides excellent illustrations. Accessible to laypersons.

Harland, David M. *Water and the Search for Life on Mars*. New York: Springer Praxis, 2005. A historical review of telescope and spacecraft observations of the Red Planet up through the Spirit and Opportunity rovers. Covers all aspects of Mars exploration, but focuses on the search for water, believed to be the most necessary ingredient for life.

Hartmann, William K. *Moons and Planets*. 5th ed. Belmont, Calif.: Thomson Brooks/Cole, 2005. An updated version of a classic text on planetary science. The material on Mars covers all aspects of ground-based and spacecraft observations of Mars. Takes a comparative planetology approach rather than presenting individual chapters on each major object in the solar system.

Hartmann, William K., and Odell Raper. *The New Mars: The Discoveries of Mariner 9*. Washington, D.C.: National Aeronautics and Space Administration, 1974. Although its information was rendered somewhat outdated by the Viking missions, this book is valuable for its detailed information about the material uncovered by Mariner 9. Includes numerous photographs; they will be useful for students of the polar caps, because Mariner 9 photographed the south polar cap steadily for four months. Some background in astronomy would be helpful.

Kargel, Jeffrey S. *Mars: A Warmer, Wetter Planet*. New York: Springer Praxis, 2004. A member of Springer Praxis's excellent Space Exploration series. The author provides a convincing case that the picture of a dry, waterless world portrayed initially by the early Mariner probes is not the Mars of today's understanding. The book takes the reader on a search for Mars's water.

Lowell, Percival H. *Mars and Its Canals*. New York: Macmillan, 1906. Lowell speculates that the polar caps are composed of carbon dioxide, a theory that regained its vogue after Mariner 4's pictures suggested an absence of water in substantial quantities in the polar caps. The book has historical interest for sophisticated readers.

Moore, Patrick. *Guide to Mars*. New York: W. W. Norton, 1977. Dated, but a reliable book on Mars and Mars exploration, this compact volume provides an excellent chapter that focuses on the Martian ice caps, plains, and deserts. The material on the Mariner and Viking missions and their results is indispensable to any serious student of the polar caps or of Mars in general. Necessary historical information is skillfully woven into the text.

Richardson, Robert S. *Exploring Mars*. New York: McGraw-Hill, 1954. It is interesting to compare this dated book with later ones that clearly indicate the enormous progress that has been made in understanding Mars. The material on the polar caps is limited but valuable. The index is extensive and accurate.

Squyres, Steve. *Roving Mars: Spirit, Opportunity, and the Exploration of the Red Planet*. New York: Hyperion, 2006. Written by the principal investigator for the Mars Exploration Rovers Spirit and Opportunity, this fascinating book provides a general audience with a behind-the-scenes look at how robotic missions to the planets are planned, funded, developed, and flown. A personal story of excitement, frustrations, a scientist's life during a mission, the satisfaction of overcoming difficulties, and the ongoing thrills of discovery.

MARS'S SATELLITES

Categories: Mars; Natural Planetary Satellites; Planets and Planetology

The two satellites of Mars, Phobos and Deimos, almost certainly originated as captured asteroids. These two small satellites could serve as future way stations for human exploration of Mars. They may provide future astronauts visiting Mars with an orbital base and even essential resources.

Overview

In early August, 1877, astronomer Asaph Hall began his search for Martian satellites at the U.S. Naval Observatory in Washington, D.C. His search was initiated for two primary reasons. He found that many astronomy texts and ephemerides of the day contained serious errors and incorrect statements. One contention was that Mars had no satellites; because none had yet been identified, that statement was correct for its time. However, Hall knew from consulting Frederick Kaiser's summary of Martian observations in the *Annals of the Leiden Observatory* (1872) that few astronomers had even looked for such bodies. While Mars made its close approach to the Earth (called "opposition") in 1877, Hall used the Naval Observatory's 26-inch Clark refractor telescope to search for potential Martian satellites.

Even as he began his search, Hall knew that the probability of finding a Martian satellite was slim. Any object even a fraction of the size of Earth's moon would have been discovered long before. Any smaller object could not even exist at any great distance from Mars, as the Sun's gravitational influence would snatch it away. Hall therefore began his search looking for a very small satellite orbiting very close to the planet, one that therefore might be very close to the visible disk of Mars as seen through the telescopic lens and thus obscured in the planet's glare. In view of these discouraging considerations, Hall said, "I might have abandoned the search had it not been for the encouragement of my wife." This statement would become indelibly etched on the discovery. On August 12, 1877, Hall first glimpsed one of Mars's two satellites, which he confirmed on August 16. The next evening, he discovered a second satellite. The announcement was made several days later by Navy admiral John Rodgers, the observatory's superintendent. These bodies were named Deimos (meaning "flight" or "panic") and Phobos ("fear") from Homer's *Iliad*: "He [Mars] spake, and summoned Fear and Flight to yoke his steed, and put his glorious armor on."

The first clear images of the satellites were made by the National Aeronautics and Space Administration's (NASA's) Mariner 9 orbiter in 1972. Five years later, even more dramatic and detailed photographs were obtained by the Viking orbiters. Those photographs revealed that the two satellites are among the darkest bodies in the solar system. Because of their density, size, and curious orbital characteristics, they are widely thought to be asteroids captured by Mars's gravitational field. They are not circular in shape. Some have even described their appearance as potato-shaped; their modest size allows for the weakest of gravitational fields, which will not permit the body to collapse into a spherical shape like that of planets and much larger satellites.

Phobos, the larger of the two Martian satellites, has a size of 28 by 23 by 20 kilometers. Phobos's orbit is exceptionally low—directly above the equator only 9,378 kilometers over the planet's surface—so low that it cannot be observed from the surface at latitudes greater than 70° north or south. The orbit of Phobos is just barely outside the Roche limit, where tidal forces would tear it apart. It probably will be torn apart and crash into the surface of Mars in the next 38 million years. Phobos's orbital period is very fast. It circles the planet in only 7 hours and 39 minutes, making the satellite appear to rise in the west and set in the east.

Deimos measures 16 by 12 by 10 kilometers. Its orbit is considerably higher than that of Phobos: 23,459

The Inner Planets

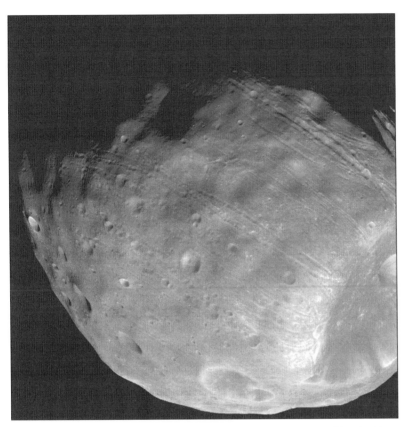

This image of Phobos, the larger of Mars's two moons, was captured by the Mars Reconnaissance Orbiter's High Resolution Imaging Science Experiment (HiRISE) on March 23, 2008. (NASA/JPL-Caltech/University of Arizona)

kilometers above the equator, with a period of 30 hours and 18 minutes. Deimos's orbit is high enough that it should eventually escape Mars's gravitational field and fly off into an independent orbit around the Sun.

Initial observations of the two satellites showed them to be heavily cratered, suggesting that their surfaces are very old—nearly as old as the solar system itself, some 4.6 billion years. There are two very large craters (relative to the body's size) on Phobos. The largest crater, dominating Phobos's northern hemisphere, is named Stickney, for discoverer Hall's wife. The astronomer himself has been remembered by a lesser surface marking: Crater Hall is a 6-kilometer depression on the satellite's southern pole. Crater Roche lies near Phobos's north pole, a reminder of the satellite's eventual fate as a result of Mars's dominant gravitational influence. As a Viking orbiter flew to within 88 kilometers of Phobos's surface, it photographed what appeared to be cracks, as wide as several hundred meters and up to 10 kilometers in length, emanating from Crater

Stickney. These cracks in the satellite's surface were almost certainly caused by the impacting body that formed the crater itself.

The surface of Deimos is not as spectacular as that of the larger satellite. A Viking orbiter flew to within 23 kilometers of Deimos and resolved a relatively quiescent surface, with smaller craters (the largest discovered is only 2.3 kilometers across), no visible cracks, and a lack of a single spectacular formation.

The theory of asteroid origins was bolstered by these visual images. An extraordinarily dark surface and a density only twice that of water (half that of Mars) lent credibility to the theory that the satellites were composed of carbon and carbon compounds such as are conjectured for type C asteroids, which populate the outermost regions of the asteroid belt. These bodies may contain up to 20 percent water by weight. Why their orbits are so nearly perfectly circular and equatorial has no easy explanation. In terms of capture probability, wayward asteroids would not necessarily slip into such neat orbits; the resulting orbit would most likely be inclined and considerably eccentric. Hence, for confirmation, such theories will have to await both the physical exploration of their surfaces and a firsthand look at carbonaceous asteroids themselves.

The Soviet Union's probes to the Martian satellite's surface were collectively called the Phobos mission. Two craft, Phobos 1 and Phobos 2, were launched in the summer of 1988; their mission was to send a lander to Phobos so that it could analyze the satellite's surface. Communication with Phobos 1 was lost before it reached Mars's orbit. Phobos 2 successfully attained orbit around Mars in 1989 and began returning photographs of Phobos and Mars. On March 27, 1989, merely days before the probe's planned close encounter with Phobos and release of its landers, the spacecraft spun irretrievably out of control, and the mission was lost.

Intense interest in the tiny satellites of Mars has been generated for several reasons. They provide a natural "space station" for future Mars explorers. Their tiny gravitational fields require very little energy to overcome, but they offer a stable platform for the staging of expeditionary landing and observation teams. They may also contain substantial quantities of water that may one day be mined to provide hydrogen and oxygen for space travelers, thus reducing the necessary burden of transporting it from Earth to the surface of Mars. Finally, less energy is required for a mission to the satellites of Mars than to and from the surface of the Earth's moon because of their weak gravitational fields and hence their low escape velocities. Such a mission, which would require a minimal round-trip travel time of two to three years, was seriously discussed by both the Soviet government and certain interests in the United States near the time when the Soviet Union collapsed in 1991. Two decades then passed with little interest in Phobos or Deimos on the part of the two major spacefaring nations.

Methods of Study

NASA spacecraft that encountered Phobos and Deimos produced photographic studies from onboard cameras and mass studies based on flyby navigational data. The most detailed photographic studies were conducted with Viking orbiter cameras. Viking 1 flew by Phobos on February 12, 1977. Viking 2 flew by Deimos on September 25, 1977. The photographic system on the orbiters was called the Viking Imaging System. It returned a total of 51,539 images of Mars and its satellites to Earth. The masses of Phobos and Deimos were estimated by determining how much the Viking spacecraft were deflected in their orbits around Mars by the gravitational fields produced by the satellites. Until the Viking encounters, the masses of Phobos and Deimos were not precisely known. Through observation of their orbital motion, coupled with a Martian mass determination, these encounters provided good estimates of the masses of Phobos and Deimos.

Photographic studies alone produced a wealth of information. A technique known as reflectivity—measuring the amount of light reflected from the surface of the satellites—enabled planetary scientists to speculate that the satellites may have been captured asteroids. It has long been speculated that a class of asteroids made of carbon and carbon compounds would be exceedingly dark, as Phobos and Deimos proved to be.

The surfaces of both satellites are saturated with craters, which indicates that their surfaces are very old. This finding enabled dating of the asteroids. In addition, the peculiar 10-kilometer crater on Phobos named Stickney displayed very large cracks down the surface of the satellite, which hinted to some planetary geologists that the composition of the body may contain substantial amounts of water ice. The cracks also indicated other subsurface structural features as well as the depth of the regolith, or top layer of soil overlying the bedrock.

Context

Phobos and Deimos may someday become two of the most important way stations in the solar system. As the Earth's focus of exploration turns to Mars as the next most logical frontier of exploration and colonization beyond the Moon, Phobos and Deimos will serve as stepping-off points to the surface of Mars. They could provide a base for operations to and from the planet. They could also provide a communications base for ground-to-space

Some Facts About Mars's Satellites

	Phobos	*Deimos*
Semimajor axis* (km)	9,378	23,459
Sidereal orbit period (days)	0.31891	1.26244
Sidereal rotation period (days)	0.31891	1.26244
Orbital inclination (deg)	1.08	1.79
Orbital eccentricity	0.0151	0.0005
Major axis radius (km)	13.4	7.5
Median axis radius (km)	11.2	6.1
Minor axis radius (km)	9.2	5.2
Mass (1015 kg)	10.6	2.4
Mean density (kg/m3)	1,900	1,750
Geometric albedo	0.07	0.08
Visual magnitude V(1,0)	+11.8	+12.89
Apparent visual magnitude (V0)	11.3	12.40

*Mean orbital distance from the center of Mars.
Source: National Space Science Data Center, NASA Goddard Space Flight Center.

and -Earth information exchanges. Finally, these two tiny satellites might be able to supply water, fuel, and oxygen to future space explorers.

Plans have been considered for sending a crewed mission to Phobos as a dress rehearsal for a later mission to the surface of Mars. This mission would test the critical life-support and medical issues that currently limit a Mars mission, at a fraction of the energy that would be required to land on the surface of the planet itself.

In the aftermath of the *Columbia* accident (February 1, 2003), the Bush administration reconsidered NASA's mission. In early 2004 the White House directed NASA to complete the International Space Station and retire the space shuttle fleet by 2010, and then to proceed with human exploration beyond low-Earth orbit. This plan called for a return to the Moon to stay and then an evolutionary approach leading to a crewed expedition to Mars. Phobos and Deimos will undoubtedly play a prominent role in eventual plans for the human exploration of Mars.

Dennis Chamberland

Further Reading

Collins, Michael. *Mission to Mars*. New York: Grove Weidenfeld, 1990. Apollo 11 astronaut Collins provides an astronaut's vision of a trip to Mars. Examines the problems to be overcome to make such a journey possible using space-shuttle-era technology.

Encrenaz, Thérèse, et al. *The Solar System*. New York: Springer, 2004. A thorough exploration of the solar system from early telescopic observations through the space missions that had investigated all planets with the exception of Pluto by the publication date. Takes an astrophysical approach, considering the solar system as just one member of similar systems throughout the universe.

Ezell, Edward, and Linda Ezell. *On Mars: Exploration of the Red Planet, 1958-1978*. NASA SP-4212. Washington, D.C.: Government Printing Office, 1984. This classic book is the National Aeronautics and Space Administration's official history of the Viking program, from the original ideas in 1958 to the culmination of the project some twenty years later. Includes discussion of the Martian moons. Generally nontechnical and accessible to all readers.

Hartmann, William K. *Moons and Planets*. 5th ed. Belmont, Calif.: Thomson Brooks/Cole, 2005. An updated version of a classic text on planetary science, covering all aspects of ground-based and spacecraft observations of Mars. Takes a comparative planetology approach rather than including separate chapters on each planet in the solar system.

Hartmann, William K., et al. *Out of the Cradle: Exploring the Frontiers Beyond Earth*. New York: Workman, 1984. This combination picture book and narrative of future human exploration discusses Mars as a logical next step for human exploration and settlement after the Moon. Depicts the potential struggles of future colonists on Mars and includes a discussion of the Martian satellites. Illustrated with an artist's conceptions of future planetary bases and explorers.

Jöels, Kerry Mark. *The Mars One Crew Manual*. New York: Ballantine Books, 1985. This speculative facsimile of a future explorer's crew manual offers an excellent reference for what the future colonist may find on arrival at Mars. Dated, but written for all backgrounds; heavily illustrated.

Kargel, Jeffrey S. *Mars: A Warmer, Wetter Planet*. New York: Springer Praxis, 2004. A member of Springer Praxis's excellent Space Exploration series. The author provides a convincing case that the picture of a dry, waterless world portrayed initially by the early Mariner probes is not the Mars of today's understanding. The book takes the reader on a search for Mars's water.

Mutch, Thomas A., et al. *The Geology of Mars*. Princeton, N.J.: Princeton University Press, 1976. A complete geological analysis of the recovered Viking data, accessible to those with a college-level science background. Illustrated with photographs and tables.

National Aeronautics and Space Administration. *The Case for Mars: Concept Development for a Mars Research Station*. San Francisco: University Press of the Pacific, 2002. This is a speculative study of how a Mars base could be established. Ironically, it came out two years before NASA was given the task of developing an evolutionary human exploration program to leave low-Earth orbit, return to the Moon to stay, and eventually travel to Mars.

Sheehan, William, and Stephen James O'Meara. *Mars: The Lure of the Red Planet*. Amherst, N.Y.: Prometheus Books, 2001. Takes a different approach to the investigation of Mars, examining what makes the Red Planet so alluring. Also describes the great astronomers who advanced humanity's understanding of Mars, from ancient times to the space age.

Zubrin, Robert. *Entering Space: Creating a Spacefaring* Civilization. New York: Tarcher, 2000. The author displays a gung-ho attitude toward making humanity a truly spacefaring species by accepting the challenge of journeying to Mars sooner rather than later, using contemporary technology and daring innovation. Speculates beyond travel to Mars.

MARS: SURFACE EXPERIMENTS

Categories: Mars; Planets and Planetology

The surface of Mars, as seen in telescopic observations and spacecraft images, hints at a history of past liquid water and possible habitability for life. Experiments conducted by robotic landers and rovers have helped scientists better understand the rocks and dust that form the planet's surface, as well as the role that water may have played in forming these materials.

Overview

While the surface of Mars has been studied extensively by telescope and spacecraft imaging (from both flybys and orbit), scientists' understanding of the surface materials and climate has been improved by data collected by robotic spacecraft located on the surface. The National Aeronautics and Space Administration (NASA) has mounted several missions, including the Viking, Mars Pathfinder, and Mars Exploration Rover (MER) missions, that have provided a context for orbital and telescopic observations and has advanced our insights into the geologic history of Mars.

The Viking mission consisted of two identical spacecraft, each with an orbiter and lander. They were launched in 1975 and arrived at Mars almost a year later. Both landers touched down in the smooth northern plains: Viking 1 in Chryse Planitia (22.3° north, 48.2° west) and Viking 2 in Utopia Planitia (47.7° north, 225.9° west). The landers contained a suite of instruments for studying the environment, a 3-meter-long robotic arm to collect surface samples for analysis, and two cameras. Images could be acquired in visible or near-infrared wavelengths in 360° around the spacecraft from 40° below to 60° above horizontal. The elemental composition of surface materials was measured using an X-ray fluorescence spectrometer and atmospheric samples were analyzed using a gas chromatograph/mass spectrometer. The latter instrument was designed to search for organic materials

A mosaic image compiled from the panoramic camera aboard the Mars Exploration Rover Spirit reveals the landscape around the rover's landing site at Gusev Crater, Mars, on January 18, 2004. (NASA/JPL/Cornell University)

as well. Further tests for evidence of life included the pyrolitic release experiment (to identify metabolism of carbon dioxide and carbon monoxide), the labeled release experiment (to detect metabolism of nutrients), and the gas exchange experiment (to examine gases released when nutrients and water vapor were added to a sample). Additional instruments included magnets (to identify magnetic components of samples), seismometers, and meteorological stations that measured wind speed and temperature.

Mars Pathfinder, launched in December, 1996, bounced onto the surface of Mars (cushioned by airbags) on July 4, 1997. Its landing site (19.3° north, 33.6° west)—which was 800 kilometers east-southeast of the Viking 1 site—was selected for safety and as part of NASA's "follow the water" strategy for Mars exploration. This location is at the mouth of Ares Vallis, a 1,500-kilometer-long outflow channel likely formed by catastrophic flooding. Rocks deposited there could have originated in any geologic unit crossed by the channel, so a variety of rocks and geologic time periods would likely be present (although source regions for the rocks would not be identifiable). The results were expected to increase knowledge of the geology and geochemistry of the Martian surface.

The Pathfinder mission consisted of a lander (named the Carl Sagan Memorial Station) and a rover (Sojourner). The Imager for Mars Pathfinder (IMP), mounted on the lander's 1-meter-high mast, provided

In the mid-1970's, the Viking 1 and 2 landers conducted the first long-term explorations of the Martian surface; this image was captured by Viking 2 on the Utopian Plain. (NASA)

panoramic stereo and multispectral imaging (up to eight wavelength bands) for rover navigation and science analysis. A set of magnets were also part of the IMP: images of the magnets were acquired throughout the mission to determine how much magnetic material from the Martian dust had been attracted to them. The final instrument system on the lander was the Atmospheric Structure Instrument/Meteorology package (ASI/MET), which measured atmospheric structure and temperature during descent and meteorological conditions (including temperature, wind speed, and wind direction) during operations.

The Sojourner rover, measuring 65 by 48 by 30 centimeters, was capable of moving up to 500 meters from the lander. The rover used cameras (two front-mounted stereo grayscale cameras, and one color camera on the rear) for navigation and science analysis. The rear camera also provided context images for chemical analyses by the Alpha Proton X-ray Spectrometer (APXS), also mounted on the rear. This instrument provided elemental compositions of rock targets (excluding hydrogen and helium) by bombarding the rock with alpha particles and measuring what type and energy of particle (alpha particle, proton, or X-ray photon) was backscattered off of the rock.

The Mars Exploration Rovers—MER A (Spirit) and MER B (Opportunity)—were launched separately in June, 2003, and landed on Mars in January, 2004. Landing sites were chosen based on engineering constraints, safety, and a continued scientific interest in evidence for past surface water. Spirit landed on the floor of Gusev crater (14.7° south, 184.5° west), a 170-kilometer-diameter impact crater that is bordered on the north by the volcano Apollinaris Patera. Gusev's southern rim is breached by a valley (Ma'adim Vallis) that terminates in the crater. The water that carved the valley may have once formed a lake in the crater. The landing site for Opportunity (2° south, 5.9° west) is located in Meridiani Planum, which was identified in orbital data as containing relatively high concentrations of the mineral hematite (Fe_2O_3).

On Earth, such concentrations are often formed by deposition in standing bodies of water.

The two MER rovers are identical in design. Each rover utilizes multiple imaging systems, including the Panoramic Cameras (PANCAMs), used for navigation and hazard avoidance, and a Miniature Thermal Emission Spectrometer (Mini-TES). The PANCAM and Mini-TES optics are mounted on the 1.5-meter-high rover mast. PANCAM provides high-resolution images in fourteen wavelengths (visible and infrared) as well as stereo coverage in any orientation around the rover (360° panorama with 180° elevation coverage). These images are used for traverse planning and scientific analysis of both surface features and atmospheric conditions. Mini-TES detects thermal infrared radiation emitted by surfaces. Because different minerals emit different amounts of this radiation, Mini-TES can be used both to measure the temperature of a surface and to identify the minerals present.

The foot of the Mars Phoenix Lander and surrounding Martian surface at the north polar region, taken on May 25, 2008. (NASA/JPL-Caltech/University of Arizona)

Each rover has a robotic arm that can extend to reach rock targets up to 0.8 meters away. The arm is equipped with a Rock Abrasion Tool (RAT) to brush away dust or expose unweathered interiors of rocks. Scientific targets are examined using the Microscopic Imager (MI). This instrument acquires images of a 31-millimeter-square area with a resolution of 30 microns per pixel. Compositional analyses are peformed using a Mössbauer spectrometer, which is sensitive to minerals containing iron, and an APXS instrument similar to the one on the Mars Pathfinder.

Knowledge Gained

The Viking landers painted a bleak picture of the Martian surface, recording temperatures from 183 to 238 kelvins (-130° to -31° Fahrenheit) and high amounts of ultraviolet radiation. Images of the landing sites show a dusty, rock-strewn surface, with seasonal formation of extremely thin water-ice frosts at the Viking 2 site. Chemical analyses of the bright red surface sediments are most consistent with iron-rich clay minerals and contain relatively high amounts of sulfur and chlorine, possibly concentrated by evaporation of water. In places, this sediment forms a thin (1- to 2-centimeter), hard crust called a duricrust. Rocks at these sites appear

dark in color, have a vesicular texture, and are likely basalt. The biological experiments proved ambiguous, as the observed results can be explained by abiotic chemical reactions. Both landers functioned for several years: Viking 1 until 1982, Viking 2 until 1980.

Mars Pathfinder images show a similar terrain, with red dust blanketing randomly strewn dark rocks, although some rocks appear to be oriented along the direction of flooding. Compositional analyses, however, suggested more than basalt. Several Pathfinder results are consistent with basaltic andesite, a more silica-rich volcanic rock. Formation of basaltic andesite requires recycling of basaltic crust, melting (and then erupting) the lighter minerals and leaving the heavier behind. On Earth, this process involves water-rich sediment and plate tectonics; the formation process for basaltic andesites on Mars is unknown. Pathfinder also provided detailed evidence of aeolian (wind) processes on Mars during its 83-sol mission, returning images of dust devils, wind streaks, and dunes.

The Mars Exploration Rovers have significantly changed our understanding of the geology of Mars. While initial analyses by Spirit found olivine-bearing basalts along the floor of the Gusev crater (unaltered volcanic rocks, not the expected lakebed deposits), areas

Martian soil on the doors of the Thermal and Evolved Gas Analyzer (TEGA), onboard the Phoenix Lander, October 7, 2008. (NASA/ JPL-Caltech/University of Arizona/Max Planck Institute)

studied later in the Columbia Hills, 3 kilometers from the landing site, included silica-rich sediment (possibly from an ancient hydrothermal system), iron sulfates, and evidence of explosive volcanism (which requires more silica-rich magma than basaltic eruptions, requiring crustal recycling).

Opportunity provided the most conclusive evidence for the presence of past surface water on Mars. Located fortuitously in a small crater, the first images showed an expanse of rock outcrop, not the scattered rocks seen at other sites. This outcrop and others examined later contain fine layers common in sedimentary rocks; some of these layers are inclined, forming crossbeds that indicate ripples or dunes. Chemical analysis identified light-colored rock in this outcrop as jarosite, an iron sulfate mineral that forms by evaporation of water. Embedded in this layer are spherules of gray hematite (dubbed "blueberries" by the science team), similar to concretions of iron oxide that form in aqueous environments on Earth.

Context

Mars has long piqued the curiosity of observers with hints of past liquid surface water. Early surface experiments suggested a surface dominated by volcanic rock (basalt), while more recent results point to a more complex geologic history with possible bodies of water, hydrothermal systems, and recycling of the basaltic crust. The Mars Exploration Rovers continue to return data that paint a more detailed geologic history of their local areas of Mars, and the Mars Phoenix lander in 2008 examined the possibility of the existence of subsurface ice and habitats for life in the Martian arctic (near 68° north). Before the arrival of winter at the Mars Phoenix landing site (near the end of calendar year 2008), a gas analysis of samples taken from the subsurface confirmed what photographs of trenches dug by the lander's robotic arm had strongly suggested, that being that indeed just centimeters below the surface water ice was present. When exposed to ambient conditions, that water ice sublimated into the Martian atmosphere. As a result of ongoing Mars exploration by robotic spacecraft in orbit and on the surface, our understanding of the Martian surface is rapidly changing.

In early 2009, NASA aired a special *Science Update* program that revealed that astronomers using a pair of telescope facilities high atop Mauna Kea in Hawaii had detected the infrared spectral signature of methane in the atmosphere of Mars. Given the nature and thinness of the Martian atmosphere, atmospheric methane could not survive the constant irradiation from solar ultraviolet rays. As a result, the implication was that this methane, found in relatively significant amounts, had to be replenished. Two production methods were possible, one geological and the other biological. On Earth, methane can be produced geologically in the conversion of iron oxide to serpentine minerals. This requires carbon dioxide, water, and an internal heat source. Mars has carbon dioxide, and thanks to the results from Mars Phoenix in 2008, it is known to have subsurface water at least in some locations. With a source of heat, serpentine mineral production on Mars might be possible. A biological origin would involve digestive processes by microorganisms. Data presented at this *Science Update* could not be used to discern whether biology or geology was responsible for the methane.

At the time of this important announcement, NASA was well into the selection process for identifying suitable landing sites for the Mars Science Laboratory (MSL), set for a 2011 launch. This discovery had the potential to shift the three-decade-old paradigm used in Mars exploration: following the water. Instead, consideration was immediately given to the search for a methane vent on the surface of the Red Planet, so that the nuclear-powered MSL might be able to conduct in situ analyses of samples and thereby determine whether Mars's methane was of geological or biological origin. If the latter proved to be true, it was likely that the primitive life-forms would subsist in subsurface layers in the presence of water and heat.

Jennifer L. Piatek and David G. Fisher

Further Reading

Beatty, J. Kelly, Carolyn Collins Petersen, and Andrew Chaikin, eds. *The New Solar System.* 4th ed. Cambridge, Mass.: Sky, 1999. Written by top scientists, these articles cover many aspects of solar system research. The Mars chapter contrasts the chemical analyses from the Viking and Mars Pathfinder missions.

Bell, Jim F., ed. *The Martian Surface.* Cambridge, England: Cambridge University Press, 2008. Compiles scientific results from 1992 to 2007 in a series of articles written by prominent Mars scientists. Provides a continuation of the last comprehensive review of Mars science by Kieffer et al. (see below). An excellent source of citations. Requires a strong background in geology, chemistry, and physics to grasp the concepts fully.

Boyce, Joseph M. *The Smithsonian Book of Mars.* Washington, D.C.: Smithsonian Institution Press, 2002. An accessible overview of Mars exploration, covering telescopic observations, mission results, the search for life on Mars, and future exploration. Places surface results from Viking and Pathfinder in context with orbital observations.

Carr, Michael H. *The Surface of Mars.* Cambridge, England: Cambridge University Press, 2005. A beautifully illustrated and citation-rich text that describes the current understanding of the Martian surface, discusses the geologic processes involved in its formation, and compares these processes to those on Earth. A similar summary for Viking results is presented in a 1981 book with the same title and author (Yale University Press).

Hartmann, William K. *A Traveler's Guide to Mars.* New York: Workman, 2003. Written in a light, engaging style, this book presents the exploration, geology, and geography of Mars in the format of a tourist's guidebook, including maps and images.

Kieffer, Hugh H., et al., eds. *Mars.* Tucson: University of Arizona Press, 1992. Review articles by top researchers provide the definitive summary of the post-Viking view of Mars. This weighty tome contains a wealth of information on nearly every topic in Mars research and is a good source for citations. A strong science background (geology, chemistry, and physics) is necessary to understand many of the concepts presented.

Raeburn, Paul, and Matt Golombek. *Uncovering the Secrets of the Red Planet.* Washington, D.C.: National Geographic Society, 1998. Written in an anecdotal style, this large-format book tells the story of Mars exploration, particularly Mars Pathfinder (author Golombek was the project scientist) from both science and engineering viewpoints. Contains many mission images, including three-dimensional anaglyphs (glasses are included) and panoramas, to illustrate the scientific results.

Sheehan, William, and Stephen James O'Meara. *Mars: The Lure of the Red Planet.* Amherst, N.Y.: Prometheus Books, 2001. Written for the general reader, this book explores the history of Mars observations from early telescopes through Viking and plans for future exploration as of the mid-1990's.

MARS'S VALLEYS

Categories: Mars; Planets and Planetology

Like Earth, Mars has valleys exhibiting complex geological histories, including flowing water, hill-slope processes, and structural control. Unlike Earth, yet similar to the Moon, Martian valleys may be as old as 4 billion years.

Overview

Valleys are low-lying, elongate troughs on planetary surfaces that are surrounded by elevated ground. On Earth, valleys often contain a stream with an outlet. About two hundred years ago, the origin of valleys on Earth was very controversial. In 1788, the Scottish naturalist James Hutton disputed the prevailing opinion that valleys formed by cataclysmic flooding, specifically the Noachian flood of biblical accounts. Hutton hypothesized that valleys formed gradually through the erosive action of the rivers and streams that lay on their floors. The fluvial origin of most valleys was subsequently demonstrated by detailed geomorphological work over the next century.

On the Moon, which lacks significant water and other volatile chemical components, valleys are very different. Valleys on the Moon are completely dry and are thought to have been created mostly by subsurface forces. Some lunar valleys may also form by chains of craters left by impacting meteors. Some, called sinuous rilles, have formed by the erosive action of lava. Others are structural depressions formed as surface blocks dropped between fractures. Such valleys also occur on Mars and Earth, but they are much less common than the fluvial forms.

In 1972, spacecraft images revealed the presence of channels and valleys on Mars that appeared very similar to those on Earth. On Mars, a very interesting inversion of scale occurs for channels and valleys. Channels are those troughs in which fluid flow once completely surrounded the depression that confined it. Martian channels are up to 200 kilometers wide and 2,000 kilometers long. The channels are much larger than networks of small valleys in the ancient, heavily cratered uplands of the planet. The valleys are typically a few kilometers wide and several hundred kilometers long. Both Martian channels and Martian valleys are extremely ancient by terrestrial standards but are comparable in age to many features on the Moon. The valleys formed early in the planet's history—by analogy to the Moon, more than 3.5 billion years ago—when rates of impact cratering were much higher than they are at present. The channels are somewhat younger, extending in age from about 3 to 0.5 billion years ago.

Martian channels were formed by immense flows of fluid that emanated from zones of collapsed topography known as chaotic terrain. This fluid seems to have burst onto the surface as immense floods of water plus considerable sediment. In the extreme cold of the Martian environment, the water would have partly frozen to form ice jams driven by the turbulent water. Local blockades of ice may have induced secondary floods, and the ice itself could have flowed in a manner somewhat similar to terrestrial glaciers. Processes of cataclysmic water outburst were probably repeated over long periods of time. These floods were probably generated from ground ice in the subsurface, perhaps heated by volcanic activity.

One of the enigmas about flowing water on Mars is the present surface environment of the planet. Surface atmospheric pressure is only about 0.7 percent of terrestrial atmospheric pressure at sea level. The temperatures measured at the Viking landing sites on the planet ranged from 243 kelvins by day to 193 kelvins at night. Subsequent landers verified the surface conditions first reported by the Vikings. Under these conditions, any standing body of water would rapidly vaporize and freeze. Rapid outbursts of water that formed the large channels, however, could have been maintained because of the relatively short duration of the flow events.

The existence of small valley networks of the heavily cratered terrains of Mars poses a problem for scientists. The valleys have short tributaries that end in abrupt valley heads, similar to the box canyons of the western United States. This valley form is believed to be caused by spring sapping. Sapping is the process whereby groundwater undercuts hill slopes; the groundwater apparently emerged as springs at the heads of the valleys, providing a subsurface source for flow.

One way to have maintained the groundwater flow that sustained Martian valley growth would have been for precipitation (rain or snow) to have fallen in the headwater areas. Water infiltrating the ground would then have recharged the groundwater flow system. This mechanism, however, requires Mars to have once had a very different climate from the one observed today. The atmosphere would have had to be warm and dense enough to hold considerable water. Atmospheric scientists have constructed theoretical models of such an ancient, hypothetical atmosphere for Mars. They conclude that increased amounts of carbon dioxide may have been present early in the planet's history. The carbon dioxide could have contributed to a greenhouse effect, whereby the planetary surface is warmed by trapping incoming solar radiation as heat. An alternative mechanism for maintaining groundwater flow to the valley networks would have been for geothermal systems to drive the flow by convective, or circulating, heat. Subsurface volcanic rocks would supply the heat, circulating the groundwater in a manner similar to that in areas of hot springs, such as Yellowstone National Park. Water flowing in the valleys would cool, seep into the ground, and recharge a recirculating system driven by volcanic heat. This mechanism would not require a dense, warm atmosphere early in the planet's history.

Ius Chasma's southern trench, near the Valles Marineris—the "Grand Canyon" of Mars. Layers of lava flows alternate with bright layers of sedimentary rock, which may be the alluvia carried by ancient Martian water flowing on the planet's surface. (NASA/JPL-Caltech/University of Arizona)

Both channels and valleys on Mars are modified by many other processes besides water flow. Because these features have hill slopes, a variety of gravity-induced slope adjustment forms are present; called mass movements, these slope adjustments include landslides, flows of debris, and slow creep of slope materials. All these processes may have been facilitated by water and ice mixed with the rock materials. Movement of debris onto the channel and valley floors, in some cases, completely conceals evidence of fluvial action that originally cut into the landscape. The Mars Global Surveyor spacecraft provided images, taken years apart, that demonstrated slumping of walls, albeit in craters, that likely was the result of water under the surface diminishing the load-bearing capability of those walls.

Wind action is also facilitated by the confinement of valleys. Wind erodes fine sediment, producing a lineated topography that parallels prevailing directions. Eroded materials may locally accumulate as sets of sand dunes, or they may be more broadly distributed as sheets of deposited sand or dust.

Valleys also served as troughs along which erupted lavas descended from volcanic source areas. Indeed, Martian volcanoes serve as excellent sites in which to observe the evolutionary sequence of valley development on Mars. Volcanoes vary in age and in the character of their surfaces, thereby providing a kind of natural experiment on the formation of valleys. Studies of fluvial valley development on volcanoes indicate that incision by flowing water occurs only when the very permeable lava flows of the volcanoes are altered to have less permeable surfaces. Lowered surface permeability arises from volcanic ash that mantles local areas. Channels forming on this ash incise into the volcano. As valleys form by enlargement of these zones, the incision is able to tap groundwater in the permeable lava flows. This groundwater further sustains valley growth in a headward direction by sapping. Eventually, the volcano is dissected by a mature network of valleys adjusted to the water flow system that is sustaining their growth.

Many mysteries still surround the valleys on Mars despite a rich history of spacecraft imaging over four decades from Mariner 4 to the Mars Reconnaissance Orbiter and Mars Express. While most valley networks are very old—older than most rocks on Earth—some are quite young. Very well developed valleys occur on the relatively young Martian volcano Alba Patera. Valleys there are restricted to a local area of volcanic ash. It may be that water was introduced by local precipitation, perhaps related to outburst flooding in the large channels.

Mars has additional surprises, including an abundance of linear grooves and a lack of depositional forms. Also puzzling is the fact that the large channels on Mars contain landforms that are very different from landforms generally seen on Earth. The best Earth analogy to the outflow channels is a region called the Channeled Scabland in Washington State. This area of flood-eroded basalt was generated by immense glacial-lake outbursts during the ice ages. The Channeled Scabland is more similar to the Martian channels than any other region on Earth.

Methods of Study

Channels and valleys of Mars were discovered by remote-sensing observations generated from spacecraft. Despite Percival Lowell's accounts of Martian "canals" nearly one hundred years ago, telescopic views of the planet were inadequate to interpret the presence of channels and valleys. It was not until 1972, when the Mariner 9 spacecraft returned the first high-resolution pictures of them from orbit about Mars, that the importance of the valleys was realized.

Ages of the valleys are interpreted by the numbers of superimposed impact craters. Then, by analogy to known cratering rates and histories on the Moon, ages can be assigned to the Martian landforms. The genesis of the valleys must also be determined by analogy. An interpreter of the pictures of the valleys must have a broad familiarity with natural landscapes, which is used to infer causes for the combinations of features seen in the channels and valleys. Often details of planetary landforms are somewhat different from what is generally known from terrestrial experience. On Mars, such lack of correspondence arises from lower gravitational acceleration, the low surface pressure, and low temperature.

Details of Martian landforms are analyzed on special maps that show relationships and patterns. Geological maps elucidate the time sequence of development, and geomorphological maps show relationships among the planetary features. Quantitative measurements can be made of the landform shapes, which can be compared to measurements on similar-appearing terrestrial features.

Model building is the activity whereby an explanation of the valley is provided in a form that extracts significant elements from the natural complexity of the phenomenon. The model may be expressed in abstract mathematical terms; it can involve laboratory hardware; or it can simulate the sequence of landform development through various kinds of analogy. In all cases, however, the model is used to express in simple, predictable terms the complexity of the real-world system under investigation. Models are only as good as their correspondence to the natural system, however, so successful model building requires an intimate knowledge of the system under investigation. This knowledge must be continually checked against new data about that system gained through ongoing investigation.

Mars studies can be separated into two distinct categories. During the initial discovery phase, Mars was visited by Mariner spacecraft. Mariner 4 flew by, snapping a small number of pictures in 1965 that revealed a Moon-like character to the surface of the Red Planet. It was something of a disappointment that the more sophisticated Mariner 6 and 7 flyby missions in 1969 verified the arid, cratered character indicated by the previous Mariner spacecraft. Then Mariner 9 orbited Mars and revealed valleys and river features as well as giant volcanoes, indicating that Mars had a dynamic past. The Viking landers examined the plains and found ambiguous soil analysis results that could not definitely answer the question about life on Mars. These missions completed the discovery phase.

Beginning with the failed Russian Phobos missions and followed by the National Aeronautics and Space Administration's (NASA's) robust sequence of spacecraft—Mars Observer (unsuccessful), Mars Pathfinder (a rover), Mars Global Surveyor, Mars Climate Orbiter (unsuccessful), Mars Polar Lander (unsuccessful), Mars Odyssey, the Mars Reconnaissance Rovers, and Mars Reconnaissance Orbiter—along with the European Space Agency's Mars Express, a second phase of focused Mars studies examined planetary features in large part seeking an answer to one fundamental question in particular. That question was essentially where Mars's water had gone.

A third phase of Mars exploration will begin once humans develop a capability to journey to Mars and complete research in situ and travel across the Martian surface with sophisticated instruments.

Context

The valleys of Mars reflect immense environmental changes that occurred on the planet. Surface conditions presently are too cold and the atmosphere too thin for water to exist in its liquid state. However, in the past, during selected epochs of its planetary history, Mars seems to have been able to sustain an active hydrological cycle that produced river valleys similar to those on Earth. Because the small valleys occur throughout the heavily cratered terrains of the planet, the scale of environmental change must have been global. In July, 2008, the Mars Phoenix lander sampled subsurface water ice near the northern polar region, providing the first direct evidence of water still on Mars.

In 2008 data from the Mars Reconnaissance Orbiter led to the conclusion that Mars of the past possessed a wet environment, something that had been suspected after the tantalizing images from Mariner 9 led scientists to alter their assessment of the planet Mars after the disappointing dry, cratered surface revealed by the Mariner 4, 6, and 7 flyby missions. Current studies strongly indicated that ancient highlands on Mars, which constitute nearly half the planet's surface, contain certain types of clay minerals that need water in order to form. This suggests that Mars once had large lakes, flowing rivers, and smaller wet environments for prolonged periods of time in the distant past. Water erosion moved the clay minerals into delta formations. The crater Jezero appears to have once confined a lake that was later breached; its water carried the clay minerals out into a fan-shaped structure.

Earth has also been affected by global environmental change. Numerous times over the past several million years, the planet has experienced decreased global temperatures with associated glaciation. During glacial advances, hydrological conditions were profoundly changed. Most recently, the global change of interest on Earth has become the warming associated with artificially increased levels of carbon dioxide and other trace gases in the atmosphere. Thus, like Mars, Earth oscillates between periods of increased warmth and cold. By comparing theories that explain such cycles, scientists hope to understand exactly how the environmental change occurs and thus to develop a means of predicting future change that will affect humanity on its home planet.

In a broad sense, the development of humankind has been linked to discoveries associated with exploration and with the migration of peoples to new lands; the space program is the most modern manifestation of such trends. When the channels and valleys of Mars were discovered, they stimulated an immense scientific effort to explain the conditions on Mars that made flowing water possible in the past. Scientists generally agree that most of the water on Mars is now locked up in its subsurface, frozen as ground ice in layers of thick permafrost. There is speculation that, if Mars could be made warm, perhaps by inducing plant growth or by releasing trapped carbon dioxide, it is not inconceivable that Martian rivers could be made to flow again—and might even sustain a population of emigrants from Earth.

Victor R. Baker

Further Reading

Atreya, S. K., J. B. Pollack, and M. S. Matthus, eds. *Origins and Evolution of Planetary and Satellite Atmospheres*. Phoenix: University of Arizona Press, 1989. A collection of research articles by planetary scientists. Provides a comparative look at Venus, Earth, and Mars.

Baker, Victor R. *The Channels of Mars*. Austin: University of Texas Press, 1982. This 198-page book provides a complete review of scientific ideas about channels and valleys on Mars. Abundantly illustrated with pictures of Martian landforms, diagrams illustrating processes, and interpretive maps. Dated, yet provides considerable discussion of possible terrestrial analogues to features on Mars. Also provides general review material on the history of Mars studies, the general geology of Mars, and implications for global environmental change on the planet.

Barlow, Nadine. *Mars: An Introduction to Its Interior, Surface, and Atmosphere*. Cambridge, England: Cambridge University Press, 2008. An interdisciplinary text including contemporary data from the Mars Exploration Rovers and Mars Express. Each chapter contains necessary background information. Excellent for both students and nonspecialists.

Carr, Michael H. *The Surface of Mars*. New Haven, Conn.: Yale University Press, 1981. A well-illustrated summary of Mars geology, this book extensively features pictures from the Viking orbiter spacecraft. One chapter covers channels and valleys, and all the important landforms of Mars are discussed in some detail. This classic 232-page book is the best starting point for an in-depth review of modern ideas about the geology of Mars.

Greeley, Ronald. *Planetary Landscapes*. Winchester, Mass.: Allen & Unwin, 1994. Provides a review of the geomorphology of planetary surfaces throughout the solar system. Compares the processes among multiple planetary bodies with an emphasis on properties of individual planets and moons. Channels and valleys on Mars are seen in the broader context of surfaces on the various planets.

Harland, David M. *Water and the Search for Life on Mars*. New York: Springer Praxis, 2005. A historical review of telescopic and spacecraft observations of the Red Planet up through the Spirit and Opportunity rovers. Covers all aspects of Mars exploration, but focuses on the search for water, believed to be the most necessary ingredient for life.

Hartmann, William K. *Moons and Planets*. 5th ed. Belmont, Calif.: Thomson Brooks/Cole, 2005. An updated version of a classic text on planetary science. The material on Mars covers all aspects of ground-based and spacecraft observations of Mars. That information appears in several chapters, as the text takes a comparative planetology approach.

Kargel, Jeffrey S. *Mars: A Warmer, Wetter Planet*. New York: Springer Praxis, 2004. A member of Springer Praxis's excellent Space Exploration series, arguing that the dry, waterless world portrayed initially by the early Mariner probes is not the Mars of today's understanding. The book takes the reader on a search for Mars's water.

Mutch, Thomas A., et al. *The Geology of Mars*. Princeton, N.J.: Princeton University Press, 1976. This classic 400-page book was produced at the beginning of the Viking mission to Mars as a summary of knowledge up to that time. Reflects the discovery of Martian channels by the Mariner 9 mission in 1972. Although many of the ideas in the book have been superseded by analysis of the Viking and later results, it provides good background on the evolution of scientific thinking about Mars. Many of the problems identified remain unresolved.

Squyres, Steve. *Roving Mars: Spirit, Opportunity, and the Exploration of the Red Planet*. New York: Hyperion, 2006. Written by the principal investigator for the Mars Exploration Rovers Spirit and Opportunity, this fascinating book provides a general audience with a behind-the-scenes look at how robotic missions to the planets are planned, funded, developed, and flown. A personal story of excitement, frustrations, a scientist's life during a mission, the satisfaction of overcoming difficulties, and the ongoing thrills of discovery.

Zubrin, Robert. *Entering Space: Creating a Spacefaring* Civilization. New York: Tarcher, 2000. The author displays a gung-ho attitude toward making humanity a truly spacefaring species by accepting the challenge of journeying to Mars sooner rather than later with contemporary technology and daring innovation. Speculates beyond travel to Mars.

MARS'S VOLCANOES

Categories: Mars; Planets and Planetology

The discovery of enormous volcanoes on Mars as a result of early images returned by the Mariner and Viking missions led to intensified study of Martian volcanic characteristics and activities. It is believed that this study will help scientists determine relationships between geological processes on Earth and Mars and develop a unified theory about the origin and evolution of the solar system.

Overview

Volcanoes on Mars are generally larger than those on Earth. In fact, one should approach the planet with the understanding that all of its major features are gigantic, including its craters, plains, valleys, volcanoes, and polar caps. That is one of Mars's overall characteristics, and the enormous size of the Martian volcanoes is typical.

Observed Martian volcanoes seem to group themselves into two distinct regions: in the Tharsis region, atop the Tharsis Dome, and in the region known as Elysium, a large topographic region of crustal upheaval. Volcanoes in other areas tend to be smaller and older than those in these two regions, although the dispersion of Martian volcanoes is not broad. Sixteen principal volcanoes have been identified in these two prime areas, both of which are mostly in the planet's northern hemisphere. According to researchers, there are very few or no volcanoes in other regions of Mars, although the reason for this absence is not completely understood. Scientists speculate that the Martian crust is much thicker, overall, than Earth's crust and that it is much more difficult for magma to punch a hole through its surface. Also, the rigid Martian crust does not allow for tectonic plate movement, such as that which occurs on Earth.

Volcanoes on Mars are created in the same manner as terrestrial volcanoes, either by eruption through a central tunnel or by eruption through side vents in the volcano's walls. There is ample evidence that both processes have been at work on Mars in its long geologic past. In the first process, material from the planet's interior pushes up, overflows, and then cools rapidly, creating an upside-down cone with a hole in the top that eventually seals, plugging the eruption tunnel. In the second instance, magma oozes through breaks in the volcano's flanks, called vents. The major characteristics of Martian volcanoes, however, are very different from those of terrestrial volcanoes.

Olympus Mons, the largest volcano in the solar system at more than half the size of Arizona, is also three times the height of Mount Everest, at 26 kilometers. The Mars Global Surveyor captured this image in 1998. (NASA/JPL/Malin Space Science Systems)

Researchers studying Mars have determined that there is no tectonic movement, or shifting of gigantic "plates" of crustal expanse, on Mars as there is on Earth. As a result, material from Mars's interior continues to erupt through the exact same tunnel over and over again, constantly building the volcano higher and wider. On Earth, on the other hand, a volcano is likely to be moved from its original spot, drifting away by the long-term motion of tectonic plates. This movement seals the magma tunnel deep inside the volcano, preventing further eruption.

The second process, eruption of magma through side vents in the volcano's flanks, comes about by the interior materials bursting through the weakest points in the sides of the volcano. This kind of process occurs on Mars as on Earth, but the result is usually quite different. On Mars, lava flowing down the sides of the volcano spills beyond the volcano's base perimeter, creating vast lava plains that completely surround the volcano. Once again, because the volcano stays in one place, repeated eruptions through the side vents cause the lava plain to continue to grow broader and broader. On Earth, the buildup is considerably slower, if it occurs at all, because the volcano is moved away from its original eruption site by tectonic movement.

Volcanoes on Mars dwarf those on Earth. The largest of the Martian volcanoes are part of a grouping of about eight near the equator in the Tharsis Bulge, or Dome. One of the four biggest is Olympus Mons, a shield volcano judged to be about 27 kilometers high and 600 kilometers in diameter. The other three lie in a nearly straight line to the east of Olympus Mons. Beginning with the northernmost, their names are Ascraeus Mons, Pavonis Mons, and Arsia Mons. Although they are not as spectacular as Olympus Mons, their sizes are still quite impressive.

Martian calderas, large craters at volcanoes' summits caused by settling of the magma in their interiors, are similarly immense. The caldera of Olympus Mons, for example, is a highly complex collection of features measuring about 3 kilometers deep and 25 kilometers across, with its walls set at a slope of about 32°. The complex is the result of repeated caldera collapse after extrusions of magma settle and stop. Calderas of Martian volcanoes are unique, at least in the inner planets, in having many circular fractures surrounding the main caldera. In addition, the entire Olympus Mons structure is surrounded by a scarp, or cliff, which at some points is 6 kilometers high. In numerous places, there is ample evidence that lava has flowed over the sides of the scarp. Scarp lava overflows are often referred to as "flow drapes" and are quite common to Martian volcanoes; they extend the main volcano structure into the surrounding lava plain, often for many hundreds of kilometers. Flow drapes at Olympus Mons extend the volcano structure at least 1,000

kilometers outward over the rigid surrounding planetary crust. Flow drapes indicate that the scarp was formed before the time of the eruption of the magma which created the lava flows, and they help geologists to determine the age of the volcano and the extent of the extrusion of magma.

Terracing is a feature that is highly pronounced on the slopes of Martian volcanoes, created by the front lines of immense lava flows churning down a volcano's sides. Multiple terracing is often seen; it results from repeated eruptions, which create well-defined blankets of lava. Olympus Mons exhibits many such terracing features. It is difficult to imagine the enormous volume and extent of the rolling lava required to produce the many examples of terracing on this volcano.

Arsia Mons rises some 16 kilometers above the Tharsis Dome with a caldera measuring 140 kilometers in diameter. The caldera is surrounded by concentric rings of hummocky topography, some of which are graben (long, linear depressions usually found between two parallel faults). The lava flow reaches at least 1,500 kilometers into the surrounding plain, often covering earlier features of the region. Images from the Mariner 9 mission, especially, disclose many lava-flow fronts and ridges where the rolling lava came to a stop. Repeated eruptions on Mars have thrown out truly copious amounts of lava, which, in turn, have formed far-reaching, relatively smooth plains adjacent to the volcano cones. These plains may extend hundreds of kilometers or, as in the case of Alba Patera, more than 1,700 kilometers out from the central cone. Although relatively flat, these plains most often show many lava-flow patterns, such as ridges and hummocks at the leading edges of flows.

The flow of lava down the flanks of a volcano can take many different forms. All these forms are present on Mars. The thickness of the lava, steepness of the slope, and presence or absence of barrier features all cause the downward spread of lava to take different forms and speeds, creating a variety of patterns. Fine-edged flows, flat-topped flows, and flow ridges can be seen in various volcano complexes. Ages of lava flows can be determined by the presence or absence of craters and other features on the volcanoes' flanks; numerous craters, for example, would indicate that the flows are relatively older and that the craters were formed after the lava flow, whereas lava that creeps into or flows over preexisting craters indicates that the flow came after the crater was created. Some ancient Martian volcanoes, such as Tyrrhena Patera, exhibit features that have been degraded by time, sometimes so much that it is impossible to determine the location of the primary volcano. Other features of some volcanoes suggest the downward flow of ash rather than of lava.

In sum, volcanoes on Mars are somewhat different from volcanoes on Earth. Martian volcanoes are significantly larger. The volcano cone is created differently, and the absence of tectonic movement allows a Martian volcano to stand over the same site for millions of years. Earth volcanoes in the process of eruption spew out lava in large amounts, yet Martian volcanoes eject lava in still larger amounts. When lava flows down the sides of a terrestrial volcano, it is affected by the landscape—streams, trees, and boulders which may deflect the flow pattern this way or that. On Mars, however, the flanks of volcanoes have no such barriers; instead, they encounter impact craters and the patterned buildup of previous lava flows. Finally, terrestrial volcanoes do not change size appreciably; those on Mars, however, keep growing bigger.

Methods of Study

Planetologists have enlisted a wide range of analytical techniques to study Martian volcanoes. The most obvious and by far most productive have been the early images of the Martian surface gained through six highly successful uncrewed missions conducted by the National Aeronautics and Space Administration (NASA) between 1965 and 1976: Mariners 4, 6, 7, and 9 and Vikings 1 and 2. These flyby, orbiting, and surface-landing missions produced a huge amount of photographic material with which geologists have begun to piece together a profile of Mars's evolution based on the study of volcanoes. The study of volcanoes is especially enlightening because it allows a view of material that originated from deep inside the planet. The success of these early missions permitted the development of new questions in the study of Mars, and led directly to more advanced spacecraft such as Mars Pathfinder, Mars Global Surveyor, Mars Observer, Mars Reconnaissance Orbiter, Mars Express, and Mars Phoenix, many of which were committed in large part to the search for water on Mars rather than volcanic processes.

By comparing Mars's topographic features with similar features on Earth, scientists can gain an enormous amount of information in a short time. By applying geometry to shadows and large-scale topography, they can determine the sizes and trace the short- and long-term development of evolutionary features. This information is critical in the case of volcanoes; lava-flow patterns, especially, are revealing about the structural evolution of volcanic sites and the surrounding terrain. Photographic evidence can be enlarged to a surprising degree, and often, extremely detailed images captured by remote-sensing technology can be studied simultaneously by scientists all over the world. Television cameras aboard the mission spacecraft gather and transmit digital image data, which are separated into shades of gray and then transmitted across space to

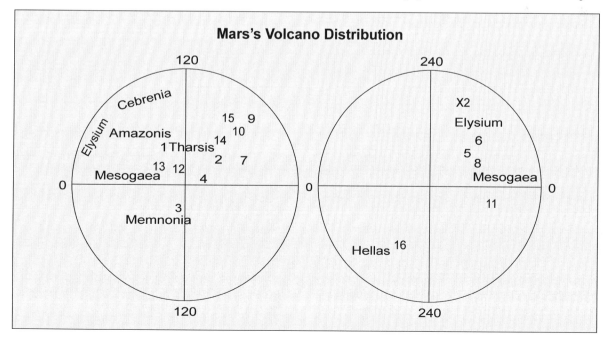

receiving antennae on Earth. In special receiving stations, digital data are put together again and stored in computers. The data can be recalled from the computers at any time and can be manipulated by computer operators. Shadow areas can be lightened or darkened, color can be added and shaded, and mathematical computations can be applied to various features, all revealing the true nature of the topography.

The Viking landers, which descended to the surface of Mars, contained a multitude of scientific instruments that recorded data about wind direction and velocity, surface and atmospheric temperatures, sizes of windblown material, composition of Mars's soil, and even viscosity of the soil. There were two landers, each settling at a different location on Mars. An arm was extended from each spacecraft, and at the end of the arm was a specially designed scoop that dug a trench in the planet's surface. By analyzing the characteristics of the trench, scientists could deduce such information as the ability of the soil to cling together, whether or not other surface material slid into the trench, and the strength it took to dig the trench. Also, a series of three instruments aboard the landers received the collected soil, analyzed it chemically, and determined its composition. Data were collected in numerical formats and radioed back to Earth, allowing scientists to create a profile of the Martian environment. The profile was then applied to other regions on the planet, such as those where volcanoes existed. The picture of Mars that developed from these instrumented explorations is one of a cold, lifeless wasteland without the kind of environment that would allow humans to survive even for a short time.

Later landers provided circumstantial evidence for water and found samples of volcanic origin locally. The next quantum leap in understanding the volcanoes of Mars would be to send geologists to the Red Planet to survey volcanic regions and perform field geology exercises in situ.

Context

Properties of Martian volcanoes are important for scientists to study because they partially reveal the planet's present and past geology, providing clues to the planet's age, formation, and activity. Considerable study of volcanoes has disclosed that there is no tectonic plate activity associated with Mars, and that the Martian crust is far more rigid than Earth's. While there is a scarcity of volcanoes on Mars in comparison with the large number on Earth, they do help to establish theories about Mars's age and evolution. Volcanic ejecta can reveal something of the activities deep within Mars's crust and of the composition of material in the planet's mantle.

If scientists could decipher how Mars evolved into the planet it is today, they could compare it to Earth and the

other planets and, perhaps, reach a greater understanding of Earth's past and future. One of the keys to unraveling Mars's mysteries is volcanic activity; it is a critical measure of what is or is not happening both on the planet's surface and deep inside its interior. Martian volcanoes are instrumental in the creation of other topographic features, such as extensive lava plains and scarps. Most volcanoes are located in a relatively confined area on Mars, and scientists want to know why volcanoes are not formed in other locations as well. Close study of Martian volcanoes could give scientists some clues about the interior properties of the planet and allow them to gain new insights into Mars's internal dynamics.

Commercially, Mars may harbor wealth in the form of natural resources. The Martian surface is characterized by planetwide iron deposits and iron oxide, from which Mars derives its pinkish-orange coloration. Other critical minerals and resources may exist there as well. It is important, therefore, to understand reasons for the present dispersion patterns of volcanoes in order to determine how volcanic actions affect mineral deposits and perhaps to predict future volcanic eruptions in mineral-rich areas.

Mars and, indeed, all planets must be viewed as large complex bodies with interacting physical systems and geological processes. Scientists, therefore, ask the following questions: What are the driving elements in the forces which sculpt, change, and characterize each planet, especially volcanic action? Is there a common thread among volcanic activities on all the planets and their satellites? How do these elements and forces behave on Earth? The study of Mars has enabled scientists to arrive at a new method known as Earth system science.

Thomas W. Becker

Further Reading

Barlow, Nadine. *Mars: An Introduction to Its Interior, Surface, and Atmosphere*. Cambridge, England: Cambridge University Press, 2008. An interdisciplinary text including contemporary data from the Mars Exploration Rovers and Mars Express. A great reference for planetary science students as well as the nonspecialist. Each chapter contains necessary background information.

Bell, Jim. *Postcards from Mars: The First Photographer on the Red Planet*. New York: Dutton Adult, 2006. An amazing number of high-quality black-and-white and color prints of Mars Exploration Rover images taken on the surface of Mars. The author was lead scientist for the rover's Pancam system; he shares the discovery process behind the photographs. Nontechnical.

Carr, Michael H. *The Surface of Mars*. Cambridge, England: Cambridge University Press, 2007. Heavily illustrated with comprehensive reference list. Author provides a complete description of the geological heating of Mars as understood based on results from Mars Global Surveyor, Mars Odyssey, Mars Reconnaissance Orbiter, Mars Express, Mars Pathfinder, and Mars Exploration Rovers.

Hartmann, William K. *Moons and Planets*. 5th ed. Belmont, Calif.: Thomson Brooks/Cole, 2005. An updated version of a classic textbook on planetary science. Material about Mars covers all aspects of the Mars exploration. Takes a comparative planetology approach rather than presenting individual chapters on each planet in the solar system.

Hartmann, William K., and Odell Raper. *The New Mars: The Discoveries of Mariner 9*. NASA SP-337. Washington, D.C.: Government Printing Office, 1974. A classic companion volume to *Viking Orbiter Views of Mars* and an overview of the startling Mariner 9 mission, this book covers basic geology and planetology, comparing Martian and terrestrial features. Includes background data about Mars and its study, an explanation of the Mariner 9 mission sequence, and interpretations of the best of the mission's visual and scientific results. Graphs, charts, drawings, and photographs. Suitable for general audiences.

Kargel, Jeffrey S. *Mars: A Warmer, Wetter Planet*. New York: Springer Praxis, 2004. A member of Springer Praxis's excellent Space Exploration series. The book takes the reader on a search for Mars's water. The author provides a convincing case that the picture of a dry, waterless world portrayed initially by the early Mariner probes is not the Mars of today's understanding.

National Aeronautics and Space Administration. *Viking 1, Early Results*. NASA SP-408. Springfield, Va.: National Technical Information Service, 1976. The first scientific results of the Viking 1 Lander are chronicled in this booklet that represents a "first look" at the instrument and photographic evidence collected by a highly successful mission. Contains many mission photographs from both the lander and the orbiter. Charts, diagrams, maps, and graphs all contribute to a valuable reference work. For the advanced high school or college student.

Sheehan, William, and Stephen James O'Meara. *Mars: The Lure of the Red Planet*. Amherst, N.Y.: Prometheus Books, 2001. This book takes a different approach to the investigation of Mars, examining what it is about the Red Planet that is found so alluring. Also describes the great astronomers who advanced humanity's understanding of Mars from ancient times to the space age.

Squyres, Steve. *Roving Mars: Spirit, Opportunity, and the Exploration of the Red Planet*. New York: Hyperion, 2006. Written by the principal investigator for the Mars Exploration Rovers Spirit and Opportunity, this fascinating book provides a general audience with a behind-the-scenes look at how robotic missions to the planets are planned, funded, developed, and flown. A personal story of excitement, frustrations, a scientist's life during a mission, the satisfaction of overcoming difficulties, and the ongoing thrills of discovery.

Zubrin, Robert. *Entering Space: Creating a Spacefaring Civilization*. New York: Tarcher, 2000. The author displays a gung-ho attitude toward making humanity a truly spacefaring species by accepting the challenge of journeying to Mars sooner rather than later with contemporary technology and daring innovation. Speculates beyond travel to Mars.

MARS'S WATER

Categories: Mars; Planets and Planetology

Knowledge of how much water there once was on Mars and how much remains would shed light on the history of Mars and the solar system, the possible development of life in Mars's past, and ways of providing resources for future Mars colonists.

Overview

There is water on Mars. How much water and exactly where it is has become the center of many questions about the planet, including several mounted by the National Aeronautics and Space Administration (NASA). In 1971, the Mariner 9 spacecraft returned photographs of Mars that clearly showed that the surface of the planet had been extensively scarred sometime in its distant past, probably by liquid water flowing over it. Subsequent detailed investigations by the Viking probes in 1976 and the Mars

The Viking 2 lander took this photo of water ice on the rocks at Utopia Planitia in May, 1979. (NASA)

Pathfinder rover in 1997 confirmed this evidence and added voluminous data to support the idea that liquid water had once existed on or close to the planet's surface.

Under present conditions, it is impossible for water to exist in liquid form on the planet's surface. The highest temperature recorded at the Viking landing sites on the warmest summer days was 244 kelvins, and the average temperature is much colder elsewhere on the planet. It fell to 150 kelvins at the site of the Viking 2 lander during Martian winter. Even if it were warm enough for liquid water to exist, it would quickly vaporize in the low atmospheric pressure of Mars. The atmospheric pressure at the Viking lander sites (the locations where the NASA probes made a soft landing on the planet's surface) was less than one one-hundredth of the Earth's atmospheric pressure at sea level. This pressure compares to that found at an altitude of about 35 kilometers above the Earth, nearly four times the altitude of Mount Everest.

Water directly measured by the Viking orbiters was in a vapor state. There was also evidence that water exists in the solid form—as ice in subsurface permafrost, in the Martian polar caps, in the surface rocks and soil, in clouds and fog, and as frost. The amount of water vapor discovered in the Martian atmosphere was quite low compared to Earth standards. At the Viking lander sites, the atmosphere was 0.03 percent water vapor. This amount compares to the percentage of water vapor at an altitude of 9 or 10 kilometers on Earth. Mars is so dry that if all the water vapor in the atmosphere of the planet could be condensed into a solid block of water ice, it would measure only about 1.3 cubic kilometers—a tiny fraction of the water vapor in the Earth's atmosphere. The Viking probes measured the part of the Martian atmosphere with the greatest amount of atmospheric water vapor, which was the area nearest the planet's north pole.

Even though the Martian air is dry by Earth standards, because of the low atmospheric pressure it is near a saturated state; daily temperature variations cause the water vapor to condense in the form of ice fogs, clouds, and frost. The fog is quite tenuous; typical Martian fogs are about half a kilometer in depth, and if they were completely condensed into water, they would form a layer only a single micron thick. Frosts of water vapor, sometimes mixed with frozen carbon dioxide to form an ice called clathrate, form on the planet's surface seasonally.

The Martian poles are repositories of water ice and frozen carbon dioxide. The depth of water ice at the north and south poles varies. The north polar cap is estimated to be one meter to one kilometer deep. The south cap is less affected by the Martian summers, and its estimated thickness is 0.23 to 0.50 meter. It is primarily solid carbon dioxide. The amount of water ice in the caps is unknown. Polar water ice directly exposed to the Martian environment can undergo a deterioration by sunlight from solid directly to vapor by a process called thermal erosion.

Estimates show that Mars's original water, outgassed from the planet's interior by volcanism (as on the Earth), may have been sufficient to cover the planet with a layer some 46 meters deep. Most of it, however, was lost over the course of the planet's history in a process called molecular dissociation. High-energy ultraviolet sunlight split the water molecules into hydrogen and oxygen atoms, which were eventually lost to space. The equivalent of nearly 270,000 liters of liquid water is lost this way each Martian day. Orbital photographs show, however, that the surface is extensively scarred by what appear to be channels whose only Earthly analogue is caused by running water. Since liquid water requires an atmospheric pressure and temperature much greater that those which now exist on Mars, it is thought that the planet's conditions must once have

been sufficiently different to allow water to flow. The Pathfinder mission supplied evidence of massive flooding.

As the orbital Viking probes circled Mars, they mapped seasonal variations of water vapor over the planet. It was discovered that as the spring and summer temperatures rose, water vapor levels increased also. Most scientists agree that the source of this water vapor was ground deposits of water. Water is encapsulated either in frozen aquifers or within the soil itself. Such frozen ground is called permafrost. Permafrost is located in the regolith of Mars; the depth of this upper layer of ice-laden soil is variable. Permafrost in the polar regions may be up to 8 kilometers deep; in the equatorial regions, it may extend to 3 kilometers.

Existence of large areas, perhaps planetwide areas, of permafrost is supported by orbital images that show four distinct geological formations: Meteoritic impacts display a kind of fluid ejecta, flowing away from the main crater, not seen on the Moon or the planet Mercury, that indicates that the energy of the meteoritic impact melted

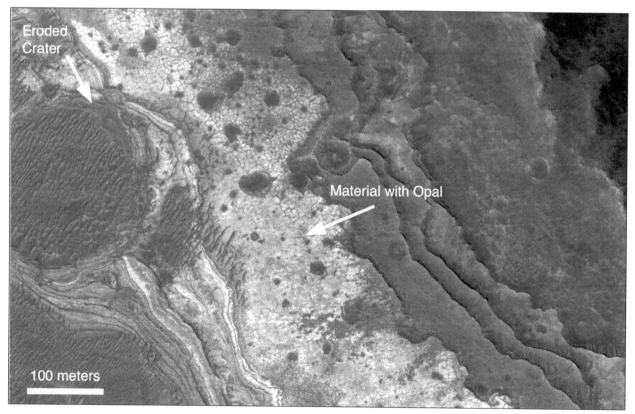

The Mars Reconnaissance Orbiter revealed hydrated rocks, similar to opal, both outside and within Mars's Valles Marineris canyons. It follows that water would have run on the surface two billion or more years ago. (NASA/JPL-Caltech/University of Arizona)

the permafrost. Another formation is called polygonal ground. These distinct polygonal shapes, observed from a high altitude, are caused by the repeated freezing and thawing of permafrost on or near the surface. Mass wasting (on Earth, often called landslides or mudslides) consists of distinctive downslope movements of soil, possibly caused by softening from water. Finally, a geological phenomenon called thermokarst was observed on Mars. It is caused by the underground melting of permafrost or underground ice formations, which lead to sinkholes or collapsed surface features.

In addition to permafrost, water is locked molecularly into the crystalline structure of the Martian soil and rocks.

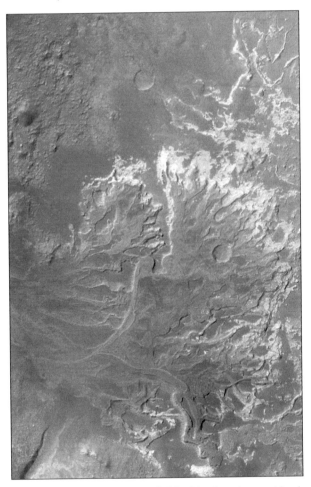

These fan-shaped, alluvial and delta-like deposits provide almost indisputable evidence of water flow on the ancient Martian surface, show that some sedimentary rocks were deposited in water, and strongly suggest that actual deltas existed where water met land. (NASA/JPL-Caltech/Malin Space Science Systems)

The Viking landers discovered that 0.1 to 1 percent of the surface materials consist of water of hydration. This tightly bound water can be released only by heating of the materials.

Many of the stream formations may have been caused by spring sapping during a planetary warming cycle. Spring sapping offers evidence of hidden, frozen deposits of ice that may thaw during warm cycles. The Mars Exploration Rovers Spirit and Opportunity provided evidence of sedimentary processes involving water. BB-sized spherules of gray hematite, referred to by the planetary scientists of the Mars Exploration Rover program fondly as "blueberries," were found by the Opportunity rover around the rim of Victoria crater. Formation of these gray hematite spherules occurs in the presence of water. Red hematite is just rust formation without the need of water, but the gray variety is found on Earth often in connection with hot springs where the oxidation of rust occurs in connection with water. However, it must be pointed out that gray hematite can be formed in connection with some types of volcanic activity, so the connection of the "blueberries" to water is not absolutely confirmed.

Methods of Study

Most of the information regarding water on Mars came from eleven NASA spacecraft: Mariner 9, two Viking orbiters, two Viking landing craft, the Mars Pathfinder lander with its small independent rover Sojourner, Mars Global Surveyor, Mars Odyssey, Mars Reconnaissance Orbiter, and the Mars Exploration Rovers Spirit and Opportunity. The European Space Agency also began its expanding planetary exploration program by studying the Red Planet from orbit with its Mars Express spacecraft.

Direct evidence for Martian water came from orbiting instruments that measured atmospheric water vapor over seasonal periods and actually photographed the Martian polar caps, fogs, and cloud formations. Landing craft directly measured water vapor on the surface of Mars. They photographed seasonal frost deposits and clouds while measuring water of hydration by heating rock and soil samples. The instrument on the orbiter that measured the water vapor was called the Mars atmospheric water detector. It examined reflected solar radiation from the Martian surface at a spectral band of 1.4 microns.

The Viking lander instrument that analyzed the Martian soil for water was called the gas chromatograph mass spectrometer. Soil samples were taken from the surface of the planet by a robotic arm that directed the sample to a heating chamber. The soil was heated to 773 kelvins, and materials driven off by the heat were analyzed. It was discovered, after a series of samples had been analyzed, that between several tenths of a percent and several percent of the surface material was water. Some of the water was believed to be loosely absorbed on the surface, but it was likely that a significant fraction was accounted for by water of hydration.

Indirectly, scientists inferred much about Martian water repositories by comparing orbital photographs with high-altitude photographs of the Earth. Nearly all investigators were convinced that the only known mechanism that could form the clearly defined river- and streambeds was running water. This observation, coupled with simplified dating techniques of counting craters in streambeds to determine approximate ages, enabled speculation about cyclic Martian warming trends. These periodic

warming trends could conceivably cause subsurface ice deposits to melt, and subsequent atmospheric pressure increases could allow the water to flow over the Martian landscape, cutting the stream formations in the soil.

Examination of ejecta patterns all over the planet led to speculation that permafrost was a planetwide manifestation. Comparison of earthly geologic formations of polygonal ground, mass wasting, and thermokarst with their Martian counterparts provided evidence for the widespread nature and even depth of the Martian permafrost layer. Photographs of cloud and fog formation over the planet during subsequent orbits enabled the calculation of temperatures, saturation levels, and even the content of cloud and fog banks. After the lander data had verified the orbital photography, a highly accurate picture of Martian water deposits was formulated.

The southern highlands of Mars, in this image from the Mars Reconnaissance Orbiter, feature braided gullies typical of water channels. (NASA/Caltech/University of Arizona)

The Mars Polar Lander was sent to touch down in a polar region of Mars and search for direct evidence of subsurface water. Unfortunately, the spacecraft crashed and failed to transmit any data. The Mars Phoenix lander was designed to attempt the same thing. Launched in August, 2007, Mars Phoenix landed in the Red Planet's northern polar region on May 25, 2008, and began to dig in the soil to search for evidence of water. Mars Phoenix accomplished the first successful powered landing on Mars since the Viking touchdowns in 1976.

Mars Phoenix touched down in the northern polar region on May 25, 2008, at 68.2° north, 234° east, a location within what scientists had named the "Green Valley" of the Vastitas Borealis region. Locally, it was late spring at the time, but the surface temperature was still sufficiently cold that solid ice permafrost was strongly anticipated. After some difficulties with the lander's robotic arm and a few other critical systems, Mars Phoenix dug a trench in the Martian soil and exposed a white layer just below the surface. In time that white layer displayed sublimation, the phase change from solid directly to a gas. The rate at which the white layer sublimated strongly suggested that it must be water ice rather than dry ice (frozen carbon dioxide, which sublimates at an even greater rate at the local Martian temperature).

Mars Phoenix was outfitted with a Thermal and Evolved Gas Analyzer (TEGA), essentially a combination of an oven and gas spectrometer. With TEGA, project scientists sought to detect water vapor released from heated samples of the white layer delivered to the hardware's oven chamber by actions of the robot arm and its scoop. Initial problems with clogging a TEGA sample inlet forced project scientists to be extremely careful in preparing for TEGA analysis. This delayed an unambiguous answer regarding whether or not the permafrost was water for several weeks, as Martian winter approached and threatened to shut down the Mars Phoenix lander.

It came as something of a relief when, on July 30, the Mars Phoenix lander's robotic arm delivered for the first time some viable subsurface material to an open chamber in the Thermal and Evolved Gas Analyzer. Several days of testing with the sticky Martian soil had produced a method whereby the arm's scoop could drop frozen permafrost into the vent leading down to a TEGA oven. The sample was heated in the oven and the evolved gases were analyzed. Just as the science team expected, the signature of water was confirmed. University of Arizona scientist William Boynton declared:

We have water! We've seen evidence for this water ice before in observations by the Mars Odyssey orbiter and in disappearing chunks observed by Phoenix last month, but this is the first time Martian water has been touched and tasted.

In the wake of this discovery, NASA announced that funds would be forthcoming to extend the Mars Phoenix mission through at least September 30, a five-week extension.

Meanwhile, NASA's sequence of orbiters—Mars Global Surveyor, Mars Odyssey, and Mars Reconnaissance Orbiter (MRO)—trained a variety of instruments on the surface of the Red Planet. Their investigations were part of NASA's continuing program directive to search for water on Mars. MRO carried the largest cameras ever flown to another planet. The MRO data indicated the presence of underground water ice, and its photographs provided circumstantial evidence for recent changes in the surface where water may have played a role.

A pair of studies using MRO came to the conclusion that Mars once had water to the extent that large lakes and dynamic rivers existed for a long time in the distant past. In the July 17, 2008, issue of *Nature*, data were presented that showed that the Red Planet's ancient highlands, essentially 50 percent of the plant, contain clay minerals that can form only with water. Those clay features were later covered by volcanic lavas. However, the clay was uncovered across the surface by subsequent impact crater events. Features like the crater Jezero once confined a lake, and clay minerals were eroded down from the crater into a delta formation. The presence of these phyllosilicate minerals across the planet added fuel to the possibility that Mars once enjoyed wet environments that might have had the potential for development of primitive life.

Context

Mars Reconnaissance Orbiter studies in 2008 concluded that Mars in the past did have a wet environment as had been originally suspected after Mariner 9

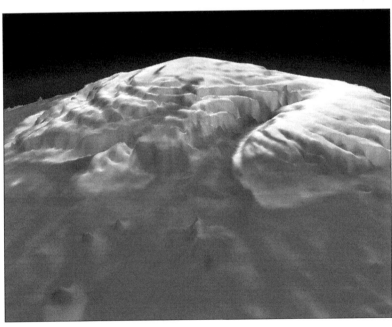

This three-dimensional image of Mars's north polar region has helped scientists determine the amount of water that may be in the ice cap. The image boasts a spatial resolution of 1 kilometer and a vertical resolution of between 5 and 30 meters. (NASA/JPL/GSFC)

images changed the scientific assessment of Mars. The newest studies suggested that Mars's ancient highlands contain clay minerals that form only with water. It appears that Mars had large lakes, vibrant rivers, and smaller wet regions that persisted for thousands and perhaps millions of years.

The question of what happened to Mars's water is of critical importance to the next generation of space explorers. It is also important to understanding how Earth's water reserves are balanced on a planetwide scale. Learning how nearly an entire planet's water resources, consisting of many trillions of liters, could simply vanish is critical not only to our insights into Mars's history but also to our understanding of Earth's future. Although the mechanisms of planetary water loss are understood, it is important to study Mars to learn exactly how and where the planet absorbed its remaining water resources. Scientists use techniques that could be employed to locate Earth's diminishing freshwater resources by space observations. Scientists may also learn how permafrost water deposits are linked to the contamination of water by soil salts and other impurities and how long-term climatic cycles lead to planetwide weather changes. If the Earth tilted only a few degrees, extreme global changes could be introduced that could have ramifications for the planet's long-term weather patterns.

Perhaps the most direct knowledge to be gained, however, is whether future colonists will be able to use what water there may be on Mars. Local water will be vital for the establishment of a Mars colony and will ultimately determine its size and usefulness. Water that is obtained from atmospheric distillation, the permafrost, the water of hydration, mining aquifers, or the polar caps will be used for a multitude of purposes, including drinking, agriculture, cooling equipment, washing and cleaning, and breaking down of molecular water into atomic hydrogen, for fuel, and oxygen, for breathing. Water on Mars may become one of the most significant aspects of "the desert planet."

Dennis Chamberland

Further Reading

Barlow, Nadine. *Mars: An Introduction to Its Interior, Surface, and Atmosphere.* Cambridge, England: Cambridge University Press, 2008. An interdisciplinary text including contemporary data from the Mars Exploration Rovers and Mars Express. A great reference for planetary science students and nonspecialists alike. Each chapter contains necessary background information.

Beatty, J. Kelly, Carolyn Collins Petersen, and Andrew Chaikin, eds. *The New Solar System.* 4th ed. Cambridge, Mass.: Sky, 1999. This beautifully illustrated and well-crafted book was intended to bring planetary discoveries to light in a single source. It covers the Martian water question in detail and extends it to a discussion of such ideas as life on Mars and the possibility of long-term Martian seasons. Written for the general reader.

Bell, Jim. *Postcards from Mars: The First Photographer on the Red Planet.* New York: Dutton Adult, 2006. An amazing number of high-quality black-and-white and color prints of Mars Exploration Rover images taken on the surface of Mars. The author was lead scientist for the rover's Pancam system. Shares the discovery process behind the photographs; not technical. Author states that his goal was to "share the beauty, desolation, grandeur, and alien strangeness" of Mars.

Carr, Michael H. *The Surface of Mars.* Cambridge, England: Cambridge University Press, 2007. Heavily illustrated with a comprehensive reference list. Author provides a complete description of the geological heating of Mars as understood based on results from Mars Global Surveyor, Mars Odyssey, Mars Reconnaissance Orbiter, Mars Express, Mars Pathfinder, and the Mars Exploration Rovers.

Collins, Michael. *Mission to Mars.* New York: Grove Weidenfeld, 1990. Apollo 11 astronaut Collins provides an astronaut's vision of a trip to Mars. Examines the problems to be overcome to make such a journey possible using space-shuttle-era technology.

Ezell, Edward Clinton, and Linda Newman Ezell. *On Mars: Exploration of the Red Planet, 1958-1978.* Washington, D.C.: Government Printing Office, 1984. This book is an official history of the Viking program, from its conception in 1958 to the culmination of the project some twenty years later. It is a detailed assessment of the program's political and technical history, but it also discusses details of the instruments that scanned for water, the Martian environment, and subsequent findings on the planet. Generally nontechnical and accessible to all readers.

Harland, David M. *Water and the Search for Life on Mars.* New York: Springer Praxis, 2005. A historical review of telescope and spacecraft observations of the Red Planet up through the Spirit and Opportunity rovers. Covers all aspects of Mars exploration, but focuses on the search for water, believed to be the most necessary ingredient for life.

Kargel, Jeffrey S. *Mars: A Warmer, Wetter Planet*. New York: Springer Praxis, 2004. A member of Springer Praxis's excellent Space Exploration series. The author provides a convincing case that the picture of a dry, waterless world portrayed initially by the early Mariner probes is not the Mars of today's understanding. The book takes the reader on a search for Mars's water.

Sheehan, William, and Stephen James O'Meara. *Mars: The Lure of the Red Planet*. Amherst, N.Y.: Prometheus Books, 2001. This book takes a different approach to the investigation of Mars, examining what it is about the Red Planet that is found so alluring. Also describes the great astronomers who advanced humanity's understanding of Mars from ancient times to the space age.

Squyres, Steve. *Roving Mars: Spirit, Opportunity, and the Exploration of the Red Planet*. New York: Hyperion, 2006. Written by the principal investigator for the Mars Exploration Rovers Spirit and Opportunity, this fascinating book provides a general audience with a behind-the-scenes look at how robotic missions to the planets are planned, funded, developed, and flown. A personal story of excitement, frustrations, a scientist's life during a mission, the satisfaction of overcoming difficulties, and the ongoing thrills of discovery.

Zubrin, Robert. *Entering Space: Creating a Spacefaring Civilization*. New York: Tarcher, 2000. The author displays a gung-ho attitude toward making humanity a truly spacefaring species by accepting the challenge of journeying to Mars sooner rather than later with contemporary technology and daring innovation. Speculates beyond travel to Mars.

MERCURY

Categories: Mercury; Planets and Planetology

Mercury, the planet closest to the Sun, superficially resembles Earth's moon. Much that was known about this planet was obtained from experiments on board and photographic images returned by the uncrewed Mariner 10 probe, which completed three flybys of Mercury in the 1970's. A new round of investigations began early in the twenty-first century with MESSENGER mission.

Overview

Mercury completes one revolution about the Sun in only 87.97 days. Mercury's orbit has a mean distance from the Sun of only 0.387 astronomical unit (1 AU is the mean Earth-Sun distance), an eccentricity of 0.206, and an inclination of 7° with respect to the ecliptic plane. Mercury rotates about an axis with no obliquity and has a period of 58.65 days. The ratio of Mercury's rotational period to its revolution period is almost precisely two to three (2:3). Mercury's mass is 3.30×10^{23} kilograms, and its mean radius is 2,439 kilometers; therefore, Mercury's mean density is 5,420 kilograms per cubic meter. Mercury's most prevalent features are craters, scarps, and deformed terrain. Degradation of original craters has resulted from secondary impact and ballistic infilling, seismic activity resulting from impacts, lava flows, and isostatic readjustment.

Mercury's surface appears remarkably similar to that of Earth's moon, although Mercury's radius is about 50 percent larger than the Moon's. Both bodies are heavily pockmarked with impact craters. Closer examination, however, reveals many important differences between the surfaces of Mercury and the Moon. The Moon has greater color variations across its surface than does Mercury. Mercury's albedo, or reflectivity, is 0.12, a brightness similar to that of the lunar highlands seen on the Moon's Earth-facing side. Although there are 20 percent albedo contrasts across Mercury's surface, it lacks the dark maria and filled craters so prevalent on the Moon. On both worlds, younger craters are often higher in albedo and surrounded by prominent ejecta blankets and bright rays. On Mercury, most craters are less than 200 kilometers across. Many of the larger craters are double-ringed with flat floors that are usually shallower than their lunar counterparts. Central peaks are found in intermediate-sized craters, but larger circular features tend not to have central peaks. Other lunar features are absent or rather rare on Mercury. There appears to be no evidence of volcanic domes, cinder cones, or lava-flow fronts on Mercury. Rilles on the planet are usually straight rather than sinuous, are quite deep, and are as wide as 6 kilometers.

Mercury is surrounded by an extremely tenuous atmosphere of helium, argon, and neon. High daytime surface temperatures coupled with a low escape velocity lead to degassing, as the average thermal kinetic energy is sufficient to permit atmospheric escape in a relatively short period of geologic time. The average planetary surface temperature is 452 kelvins. However, the maximum dayside temperature is as much as 700 kelvins at closest approach to the Sun, and the minimum nightside temperature is 90 kelvins. Mercury exhibits the greatest equatorial temperature variation, more than 600 kelvins, of any planet in the solar system.

Mercury has a magnetic field only 1.6 percent as strong as Earth's. The origin of Mercury's magnetic field remains uncertain. A metallic core composed primarily of iron would be consistent with both the observed density and the magnetic field; however, Mercury's rotational speed could be too slow to generate currents in the core even if it is molten. Mercury's small magnetic field interacts with the solar wind. Mariner 10 recorded a moderately strong bow shock that traps energetic solar wind particles.

Although Mercury's surface resembles the Moon's, its mean density is closer to that of Earth. It is believed that, in proportion to its size, Mercury contains, in its core, double the amount of iron found in any other world in the solar system. High iron content would be consistent with the formation of Mercury by condensation of the solar nebula close to the Sun; heavy elements would have been more attracted to the early proto-Sun than would lighter elements. Existence of a magnetic field suggests that

Mercury's iron has undergone differentiation and formed a hot, convecting core. Formation of the iron core would have generated heat in addition to the original radiogenic heat, causing expansion and melting of the entire mantle. This process would have had to occur before the era of intense cratering, because Mercury lacks surface expansion features and numerous preserved lava-flow formations.

According to one model of Mercury's interior, the planet's asthenosphere cooled quickly and thickened, possibly disappearing altogether. Mercury's lithosphere could extend down to the iron core, hundreds of kilometers below the surface. There is evidence to support such a model. After core formation, the planet would have cooled and contracted, resulting in a decrease in radius perhaps as large as one or two kilometers. Compression of the surface would then have caused thrust faults, the result of one rock unit slipping over another. Observed thrust faults on Mercury indicate a two-kilometer contraction.

Mercury's surface physiography can be classified into four major terrain types: heavily cratered terrain, smooth plains, intercrater plains, and hilly and lineated terrain. Intercrater plains are believed to be the oldest, predating the era of intense impact cratering. Smooth plains represent the youngest.

Smooth plains, located principally in the northern hemisphere near the large feature called Caloris Basin, are flat, lightly cratered surfaces akin to lunar maria. Craters in the smooth plains are typically sharp rimmed and only 10 kilometers across, at most. Some smooth plains fill the floors of large craters. Often, plains have sinuous ridges, an aspect shared by lunar maria. Regardless of location on the surface, Mercury's smooth plains have equal impact crater frequency, which indicates that smooth plain features were all formed at about the same time. Smooth plain features are believed to be volcanic in origin. There are too many smooth plains for them to have resulted from a single catastrophic impact or to be ejecta from the large Caloris Basin. Similarities between lunar maria and smooth plains suggest a common origin. Lunar maria are

Mercury, in this October, 2008, image sent back from MESSENGER, sporting massive impact craters and a series of rays from north to south. (NASA/Johns Hopkins University Applied Physics Laboratory/Carnegie Institution of Washington)

volcanic in nature, so smooth plains on Mercury are believed to be volcanic, also.

Intercrater plains, believed to be the oldest material on Mercury, form the largest physiographic feature on the planet. These plains have a greater crater density than smooth plains. Craters that pockmark rolling plains are typically less than 10 kilometers in diameter, and generally represent secondary rather than primary impacts. Intercrater plains were formed by a variety of different events occurring over long periods. These plains are probably primordial crust that has been subjected to impact cratering, but their origin is by no means clear. The variety of intercrater plains suggests several alternative origins.

Heavily cratered terrain is reminiscent of the lunar highlands, being areas of many overlapping craters. Crater diameters vary between 30 and 200 kilometers. Ejecta deposits cannot be clearly identified with individual craters because of high degrees of overlap and disruption. This variety of terrain was formed as the era of intense bombardment was ending.

Hilly and lineated terrain is found directly opposite the Caloris Basin on Mercury and may have been formed by the Caloris impact event itself. This heavily deformed terrain, often referred to simply as Weird Terrain, covers about 250,000 square kilometers. It is made of hummocky

hills 5 to 10 kilometers wide at the base and 0.1 to 1.8 kilometers high. Seismic energy from the Caloris impact apparently underwent antipodal focusing through the planet's core and broke or jumbled this region into hills and depressions.

The Caloris Basin is the largest single surface feature revealed by Mariner 10 photographs. More than 1,300 kilometers across, Caloris resembles the Moon's Imbrium Basin. It may represent an important event in Mercury's history, just as the Imbrium Basin does for the Moon. The basin is rimmed by mountains 30 to 50 kilometers wide and several kilometers high. Inside the basin are smooth plains scarred by small craters, ridges, and grooves indicative of lava flows modified by tectonic activity. Seeing the entirety of the Caloris Basin was a high-priority early objective of Mercury Surface Space Environment, Geochemistry, and Ranging (MESSENGER) flybys prior to beginning prolonged orbital studies.

Methods of Study

Mercury is a planet known to the ancients. However, Mercury reveals few of its secrets to visual observation from Earth. Because it is so close to the Sun, it is often hidden by solar glare and can be seen only briefly, visually or telescopically, at twilight or daybreak. The Hubble Space Telescope could not be used to obtain high-resolution images of Mercury's surface because of the tremendous brightness of the Sun, which would overpower and damage that orbiting observatory's sensitive instruments. Nevertheless, Mercury studies have advanced greatly since the first days of visual observations of this innermost planet in our solar system. Astronomers once incorrectly assumed that Mercury did not rotate as it revolved around the Sun. Few surface features were known before the Mariner 10 encounters. Indeed, for all intents and purposes, almost all that was known about Mercury prior to 2008 was obtained through scientific investigations performed by the Mariner 10 spacecraft on its three brief flybys in the mid-1970's. That probe was equipped with seven primary experiments; they involved high-resolution television imaging, infrared radio-metry, radio wave propagation, extreme ultraviolet spectroscopy, magnetometry, plasma detection, and charged particle flux measurements.

Mercury Compared with Earth

Parameter	Mercury	Earth
Mass (10^{24} kg)	0.3302	5.9742
Volume (10^{10} km³)	6.083	108.321
Equatorial radius (km)	2,439.72	6,378.1
Ellipticity (oblateness)	0.0000	0.00335
Mean density (kg/m³)	5,427	5,515
Surface gravity (m/s²)	3.70	9.80
Surface temperature (Celsius)	−170 to +390	−88 to +48
Satellites	0	1
Mean distance from Sun millions of km (miles)	58 (36)	150 (93)
Rotational period (hrs)	1,407.6	23.93
Orbital period (days)	88	365.25

Source: National Space Science Data Center, NASA/Goddard Space Flight Center.

The MESSENGER 2008 second flyby produced this image of a previously unknown surface region, with both impact craters and tectonically smoothed terrain. (NASA/ Johns Hopkins University Applied Physics Laboratory/Carnegie Institution of Washington)

Mercury's atmosphere was studied during a solar occultation using Mariner 10's ultraviolet experiment. The instrument measured the drop in the intensity of solar ultraviolet radiation as Mercury's disk and tenuous atmosphere obscured it. Data provided a profile of atmospheric concentration above the planet's surface. Other atmospheric data were gathered by monitoring radio waves emitted by Mariner 10 as it passed behind Mercury and then reemerged. The infrared radiometer, fixed to the spacecraft body on the sunlit side, had apertures which shielded the detectors from direct solar radiation. This experiment determined Venusian cloud temperatures as well as measuring surface temperatures on Mercury. Heat-loss data obtained as Mariner 10 crossed the planet's terminator, the line separating daylight from darkness, helped scientists infer information about the planet's surface composition. Surface brightness temperature was measured in a pair of spectral ranges, 34 to 55 micrometers and 7.5 to 14 micrometers, which represented temperatures of 80 to 340 kelvins and 200 to 700 kelvins, respectively.

Mariner 10 measured Mercury's magnetic field with a magnetometer package consisting of two three-axis sensors placed at different spots on a 6-meter-long boom. The use of two sensors provided the capability to isolate the spacecraft's own magnetic field from the weak field of the planet. Magnetic field measurements in interplanetary space were also made.

High-resolution images of any planetary surface can provide a wealth of information concerning that planet's past, its present geologic activity, and its surface composition. Mariner 10's television imaging system included two vidicon cameras attached to telescopes. The assembly was mounted on a scan platform that permitted the horizontal and vertical movements necessary for precise pointing. Cassegrain telescope systems were used in the imaging system. Powerful enough to resolve ordinary print at a distance of more than 400 meters, this system provided narrow-angle, high-resolution images. The television system also included an auxiliary optical system to obtain wide-angle, lower-resolution photography. This system was mounted on each of the television cameras. Experimenters were able to switch from narrow-angle to wide-angle imaging by moving the position of a mirror on the system's filter wheel. The vidicon cameras had 9.8-by-12.3-millimeter apertures and could make exposures of between 3 milliseconds and 12 seconds. Analog signals from the vidicon camera readout were digitized for transmission to receiving stations on Earth. An individual television image consisted of 700 vidicon scan lines, with each scan line consisting of 832 pixels.

The principal objectives of Mariner 10's television imaging program included collection of data useful in studying Mercury's planetary physiography, making a precise determination of Mercury's radius and rotation rate, evaluating Mercury's photometric properties, and categorizing the morphology of surface features. Television scans were made of the space surrounding Mercury in an attempt to locate unknown satellites, but none was found. This system was also used for studies of

Venus and Comet Kohoutek before Mariner 10 even arrived near Mercury.

Data from Earth-based radar investigations of Mercury strongly suggest that at least part of the planet's core could presently be molten. Such a molten layer would have large implications for the production of the planet's global magnetic field and variations in Mercury's spin rate if the liquid core is decoupled from the solid mantle. In a 1992 issue of *Science*, Martin A. Slade et al. presented the results of two studies using the Arecibo radio telescope, the Very Large Array, and the Goldstone tracking antenna to send radio waves to Mercury at selected frequencies and detect the reflected signals. Essentially radar-astronomy exercises, the aim of both studies was to generate a radar reflectivity map of Mercury's surface at a resolution of about 15 kilometers. In the process, radar-bright returns that were highly depolarized were encountered near the planet's north and south poles. Data suggested the totally unexpected presence of ice on Mercury. Ice very effectively reflects radar at the gigahertz frequencies used in these studies and depolarizes those reflected radio waves greatly. Some of the radar-bright areas detected in these studies coincided with crater-sized spots. This provided evidence for the supposition that ice existed in crater areas that were permanently shadowed from solar radiation and therefore not heated tremendously, as was the rest of the planet when under daylight conditions. One of the bigger radar-bright areas was the large crater Chao Meng-Fu at Mercury's south pole. Planetary scientists supposed that the proposed ice came from either meteoritic bombardment or planetary outgassing (or both). Confirmation of this surprising result would have to await MESSENGER studies.

MESSENGER carried seven instruments that produced a great deal of data during the spacecraft's January 14 and October 6, 2008, encounters—information that would take longer to analyze than the time to the next and final flyby before eventual orbital insertion in 2011. The mission's objectives included photographing the as-yet-unseen 50 percent of the planet's surface,

determining the composition and structure of Mercury's crust, understanding more about the planet's geological history, examining the planet's thin atmosphere, measuring the planet's quite active magnetosphere, searching for water at the poles, and providing data that would reveal the nature of Mercury's large core. This first attempt in thirty-three years to examine Mercury from close range was designed to help answer five separate major questions: What is the elemental and mineralogical composition of the surface? What does the surface look like at a resolution of better than hundreds of meters? What is the structure and temporal variation of Mercury's magnetic field? Does the planet's gravitational field exhibit any anomalies that might shed light on any uneven distribution of mass within Mercury? and What neutral particles and ions are found in Mercury's magnetosphere?

MESSENGER was outfitted with wide- and narrow-angle color and black-and-white imaging systems, a laser altimeter, a radio science experiment, and four multipurpose spectrometers. The spectrometers were

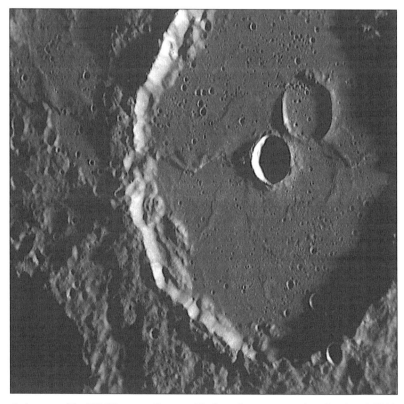

A close-up of the 106-kilometer-diameter Machaut Crater acquired on October 6, 2008, by MESSENGER. (NASA/Johns Hopkins University Applied Physics Laboratory/ Carnegie Institution of Washington)

capable of measuring the spectra of gamma rays, neutrons, energetic particles and plasmas, and the reflected light from Mercury's atmosphere and surface for compositional studies. The spacecraft's laser altimeter was designed to determine the elevation of the planet's surface features, as well as to look for wobble in the planet about its rotational axis. That sort of motion could help verify the existence of a suspected liquid layer in the core. The neutron spectrometer was designed to detect water ice at the polar regions. The laser altimeter was designed to measure the topography of permanently shadowed craters that might shelter water ice deposits. The ultraviolet spectrometer was designed to look for sulfur or hydroxyl deposits atop the water ice.

In May, 2008, researchers published results of laboratory modeling of Mercury's core that included a separated molten layer surrounding a solid core. University of Illinois and Case Western Reserve University scientists hypothesized that deep within the planet an iron "snow" forms and moves down toward the solid core. Convection could be set up and create the planet's magnetic field. This experiment investigated the behavior of an iron-sulfur sample under tremendous pressure and heat. The iron-sulfur sample was set up to mimic the suspected core structure of Mercury. If formed, molten iron condenses to flake-like crystals, which would fall to the core. This heavy iron "snowfall" would result in lighter liquid sulfur rising, establishing convection currents. Observational data for MESSENGER will be able to determine if this laboratory model actually matches Mercury's internal structure.

In the meantime other scientists at Virginia Tech reported results of different simulations of conditions on Mercury. This work suggests that the shrinking of the planet's crust as Mercury cools over geological time should produce the thrust faults seen as scallop-edged cliffs and scarps on the planet. MESSENGER will also shed light on mantle convection, a process considerably different from that on Venus and Earth because of the thinness of Mercury's crust.

Details about the prolonged analysis of MESSENGER data collected during its first flyby surfaced in science journals in early July, 2008. The data confirmed that volcanic activity had played a tremendous role in the formation of Mercury's surface, especially during a period lasting from 3 to 4 billion years ago. MESSENGER provided evidence of volcanic vents along the margins of the Caloris Basin. Other evidence demonstrated that effusion had occurred. This process sees molten material from below the crust exude upward and outward across a planet's surface, sometimes forming features that resemble volcanic shields. Mercury had suffered lava floods that filled in fairly large craters almost to the wrinkled scarps outlining the craters. Some layers of lava were determined to be as deep as 2.7 kilometers.

Context

Mercury is the Roman name for the Greek god Hermes, patron of trade, travel, and thieves. Timocharis is considered to have registered the first recorded observation of Mercury, in 265 B.C.E. Very little more was learned about the planet until the invention of the telescope. Observation of the phases of Mercury was first reported in 1639 C.E. by Italian astronomer Giovanni Battista Zupus. Telescope technology improved, and evidence of surface features was found in the early 1800's, when astronomers Karl Ludwig Harding and Johann Schröter measured albedo variations.

It was not until the early 1960's that Mercury's rotation rate was precisely measured using radar observations. Then came the launch of Mariner 10, the final spacecraft in the historic Mariner series, on November 3, 1973, at 12:45 A.M. eastern time atop an Atlas-Centaur launch vehicle from Launch Complex 36B at Cape Canaveral. Photographs obtained during this flyby mission began the geologic analysis of Mercury. This spacecraft became the first to use gravity assists from large solar system bodies to redirect its trajectory to multiple photographic targets. It was recognized that the alignment of Earth, Venus, and Mercury was such that a single spacecraft could be launched between 1970 and 1973 from Earth toward Venus and then reach Mercury. Giuseppe Colombo of the Institute of Applied Mechanics in Padua, Italy, noted during an early 1970 Jet Propulsion Laboratory conference on the approved Mariner 10 mission that a 1973 launch opportunity existed in which the spacecraft could enter an orbit with a period nearly twice that of Mercury. That meant that a second Mercurian encounter was possible. Mariner 10 was indeed placed on a trajectory that permitted multiple encounters with Mercury, and this success demonstrated the feasibility of gravity-assist trajectories. The technique would prove tremendously valuable to the Voyager probes, which were sent to the outer solar system.

Shortly after Mariner 10's escape from Earth orbit, its planetary science experiments were activated to verify their operating condition. Mosaic photographs returned to Earth indicated that the spacecraft was in good condition to image a Moon-like world with high-quality camera systems. Mariner 10 came within 5,794 kilometers of Venus on February 5, 1974. During eight days of photography, the spacecraft returned 4,165 images of Venus and a wealth of data about the Venusian atmosphere. After another forty-five days of interplanetary cruising, the spacecraft reached the mission's principal target: the planet Mercury. Mariner 10 began taking photographs on March 23, 1974, reaching its closest approach, 5,790 kilometers, on March 29. The spacecraft then passed behind Mercury, to the nightside. More than two thousand photographs were obtained on this first encounter. Mariner 10's trajectory returned the spacecraft to Mercury on September 21, 1974; this time, it came as close as 50,000 kilometers. The probe completed a third encounter in March, 1975, before running out of fuel and entering a solar orbit.

The MESSENGER orbiter was designed to continue the scientific exploration of Mercury where Mariner 10 left off more than thirty years earlier. MESSENGER launched on August 3, 2004, and was injected into an interplanetary orbit that brought it back to Earth a year later for a gravity assist that would slow the spacecraft down to fall into the inner solar system. It was directed to encounter Venus in October, 2006, and again in June, 2007, for gravity assists that set it up for its first Mercury flyby. MESSENGER flew by Mercury near its equator on its first encounter, which took place January 14, 2008. Thus, little useful information about polar ice deposits was obtained. However, a highlight of the encounter was the capture of detailed images of the remainder of the Caloris Basin, not seen during Mariner 10's flybys. All spacecraft instruments functioned, signaling the potential for prolonged study once MESSENGER attained orbit. MESSENGER's flight path was refined by thruster firings so that it again encountered Mercury, flying by on October 6, 2008, collecting data and images and also setting itself up by a gravity assist in such a way that it would fly by Mercury one more time in 2009 before eventually entering orbit about the planet in 2011.

Even before entering orbit, as a result of the three flyby encounters MESSENGER was expected to give planetary scientists a nearly full initial map of Mercury's globe. To achieve orbit, MESSENGER's main propulsion system would fire to slow down by 860 meters per second. This fourteen-minute-long burn would consume 30 percent of the spacecraft's total fuel load. The first orbit would be adjusted until MESSENGER assumed an elliptical orbit ranging from 200 kilometers to 15,193 kilometers inclined 80° to Mercury's equator. In this nearly polar orbit, the spacecraft would orbit Mercury every twelve hours. Once in orbit about Mercury in March, 2011, MESSENGER's primary mission was to last four Mercurian years (the equivalent of one Earth year or two Mercurian solar days).

The European Space Agency (ESA) and Japanese Aerospace Exploration Agency (JAXA) also plan to investigate Mercury. Both space agencies intend to launch small probes in 2013. If successful, these will reach Mercury orbit in 2019. The ESA probe is named Bepi Colombo after the late Italian mathematician/engineer Guiseppi (Bepi) Colombo, father of the gravity assist. JAXA plans to send two more spacecraft to Mercury: one designed to conduct mapping operations, and another to investigate Mercury's magnetosphere. Although not designed to last that long, MESSENGER might exceed its mission life and be able to coordinate with these other spacecraft in Mercury studies.

David G. Fisher

Further Reading

Balogh, André, Leonid Ksanfomality, and Rudolf von Steiger, eds. *Mercury*. New York: Springer, 2008. This work provides background information and reviews changes in humanity's understanding about Mercury since the Mariner 10 flybys.

Beatty, J. Kelly, Carolyn Collins Petersen, and Andrew Chaikin, eds. *The New Solar System*. 4th ed. Cambridge, Mass.: Sky, 1999. Amply illustrated with color images, diagrams, and informative tables, this book is aimed at a popular audience but can also be useful to specialists. Contains an appendix with planetary data tables, a bibliography for each chapter, planetary maps, and an index.

Clark, Pamela. *Dynamic Planet: Mercury in the Context of Its Environment*. New York: Springer, 2007. Written by a NASA space scientist who edits the *Mercury Messenger* newsletter, this book covers the search for understanding of the solar system's closest planet to the Sun.

Davies, Merton E., Stephen E. Dwornik, Donald E. Gault, and Robert G. Strom. *Atlas of Mercury*. NASA SP-423. Washington, D.C.: National Aeronautics and Space Administration, Scientific and Technical Information Office, 1978. Provides an excellent contemporary description of Mariner 10 and its mission. Includes a full atlas of spacecraft photography of Mercury. An essential reference for the planetary science enthusiast or researcher; also accessible to general audiences.

Domingue, D. L., and C. T. Russell, eds. *The MESSENGER Mission to Mercury*. New York: Springer, 2008. A compilation of articles by experts in the MESSENGER mission to Mercury. Covers the flight, the science, the spacecraft, and engineering operations needed to conduct three flybys and insert MESSENGER into orbit around Mercury. For both the planetary science student and the astronomy enthusiast.

Dunne, James A., and Eric Burgess. *The Voyage of Mariner 10: Mission to Venus and Mercury*. NASA SP-424. Washington, D.C.: National Aeronautics and Space Administration, Scientific and Technical Information Office, 1978. This book offers an elegant description of the first spacecraft mission directed to the planet Mercury. Prepared immediately in the aftermath of the mission by the Jet Propulsion Laboratory, the text is complete with information on spacecraft operations and data returns. Photographs of Mercury abound. For general audiences.

Encrenaz, Thérèse, et al. *The Solar System*. New York: Springer, 2004. A thorough exploration of the solar system from early telescopic observations through the space missions that had investigated all planets with the exception of Pluto by the publication date. Takes an astrophysical approach to give our solar system a wider context as just one member of similar systems throughout the universe.

Moore, Patrick. *Moore on Mercury: The Planet and the Missions*. New York: Springer, 2006. An astronomical survey of humanity's knowledge of Mercury. Discusses what Mariner 10 observed and outlines the MESSENGER mission.

Spangenburg, Ray. *A Look at Mercury*. New York: Franklin Watts, 2003. Covers the search for understanding about the planet Mercury from antiquity through the Mariner 10 flybys. Previews the MESSENGER mission.

Strom, Robert G. *Mercury: The Elusive Planet*. Cambridge, England: Cambridge University Press, 1987. Dated, but covers all aspects of Mercury as understood from Mariner 10 observations. Suitable for the general audience and younger readers.

Strom, Robert G., and Ann L. Sprague. *Exploring Mercury: The Iron Planet*. New York: Springer, 2004. According to a review by *Sky and Telescope*, this work is a "comprehensive text" that covers all that is known about Mercury. Contains a CD-ROM featuring Mariner 10 images and describes the anticipated MESSENGER mission.

Tanton, Linda Elkins. *The Sun, Mercury, and Venus*. New York: Chelsea House, 2006. A look at the innermost portion of the solar system and the star, our Sun, which plays such a prominent role in the evolution of both planets. For the general audience with an interest in science.

PLANETARY ORBITS

Category: Planets and Planetology

Planets in the solar system revolve around the Sun in elliptical orbits at speeds that vary with distance from the Sun. Laws that govern these motions were first deduced by Johannes Kepler and later quantified by Sir Isaac Newton.

Overview

Planets in the solar system move around the Sun in elliptical orbits. Those whose orbits are closest to the Sun move more rapidly than those that are farther away. These simple, universally accepted observations form the basis of the knowledge of planetary motions. Gravity—the force that causes apples to fall from trees and keeps humans firmly planted on Earth's surface—plays the central role in the mechanics of planetary motions.

A simple experiment illustrates the energy relationships inherent in orbiting bodies. If a person attaches a string to a small rubber ball and the ball is swung around the person's head in a horizontal circle, the tension along the string that holds the ball in its "orbit" is analogous to the Sun's gravity pulling on a bound planet. The English astronomer and mathematician Sir Isaac Newton (1642-1727) explained how the force of gravity affects planetary motion. Newton proved in his laws of motion that once an object is in a straight-line motion, it will continue on that course with no further input of energy (law of inertia) unless its motion is perturbed by an unbalanced force. In the case of planets, this force is provided by the gravitational attraction of the Sun (or more massive planet, in the case of a satellite). Depending upon the magnitude of the orbiting body's "kinetic energy" (energy of motion), the body will move in either a circular orbit or, with greater kinetic energy, an elliptical orbit. Kinetic energy counters the attractive force of gravity, thus preventing a planet from falling into the Sun, or the orbiting ball, as stated in the example, from striking the experimenter.

The scientist who first showed that the orbits of the planets are actually ellipses rather than circles was Johannes Kepler (1571-1630). A German mathematician, astronomer, and astrologer, Kepler worked previously as an assistant to the Danish observational astronomer Tycho Brahe. After Brahe's death, Kepler used detailed position measurements of the planet Mars to plot an orbit that was not circular. Up to this time, planetary orbits—including that of the Moon—were believed to be circular in accordance with precepts developed by the Greek philosopher Aristotle.

A circle is the locus of points all the same distance from a given center. An ellipse differs from a circle in being oval-shaped. An ellipse contains two internal, evenly spaced points called foci. It is important to understand how the foci of an elliptical orbit relate to the positions of an orbiting planet and the Sun. This relationship is expressed by Kepler's first law: Each planet moves around the Sun in an orbit whose shape is that of an ellipse, with the Sun at one focus. The other focus is empty. Thus, the

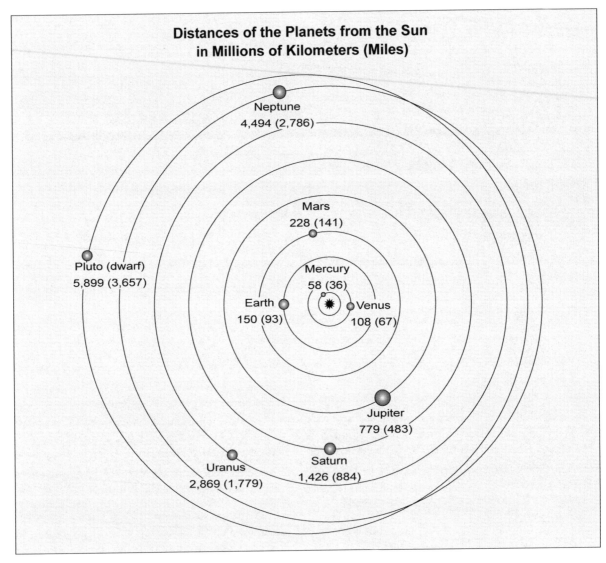

Distances of the Planets from the Sun in Millions of Kilometers (Miles)

Neptune
4,494 (2,786)

Mars
228 (141)

Pluto (dwarf)
5,899 (3,657)

Mercury
58 (36)

Earth
150 (93)

Venus
108 (67)

Jupiter
779 (483)

Uranus
2,869 (1,779)

Saturn
1,426 (884)

Sun is not precisely in the middle of the ellipse but displaced somewhat to the side. The degree of displacement determines the orbit's eccentricity. As a result, planets move between a minimum distance from the Sun in their orbit, called perihelion, and a maximum distance from the Sun, called aphelion. Planetary orbits have this repetitive pattern as a result of the central character of gravity; that is, gravity acts along the line between the gravitationally interacting bodies. The magnitude of the gravitational force follows an inverse square law with regard to its dependence upon distance between the interacting masses. If one doubles the distance between the two objects, their gravitational attraction diminishes not by a factor of two, but by four.

Kepler's second law was actually discovered before his first law. It describes the fact that planets move more slowly when farther away from the Sun (their slowest speed is at aphelion), and they move more rapidly when closer to the Sun (their maximum speed is at perihelion). This observation logically would lend support to the idea that the planet's orbit is anything but circular. The second law states that a straight line joining the planet and the Sun sweeps out equal areas in space in equal intervals of time. Imagine a string attached to a planet at one end and the Sun at the other. When the planet is near aphelion (farthest from the Sun), it moves slowly, so that the triangular sector swept out by the string during a given time will resemble a long, slender piece of pie. In contrast, near perihelion over the same time period, the planet will move farther (because it is going faster), so that the sector swept out by the string resembles a fatter slice of pie. Kepler's second law states that these two pie slices, or triangular sectors—although quite different in radius and opening angle—should have equal areas. This exercise is a mathematical way of stating that planets move more slowly as the Sun-planet distance increases. Planetary orbits obey Kepler's second law of motion as a consequence of conservation of angular momentum.

Kepler's third law, formulated in 1619, was an attempt to quantify the fact that a planet moves more slowly the farther its orbit is from the Sun. His task was to determine a precise mathematical relationship between a planet's average distance from the Sun and its period. Being oval, ellipses have a major axis and a minor axis of different lengths. A line passing through the two foci of the ellipse and ending at both ends of the figure defines the long axis of the ellipse and is known as the major axis. A length equal to one-half the major axis is called the semimajor axis. A line perpendicular

to the major axis passing halfway between the two foci of the ellipse is the minor axis. A length equal to one-half the minor axis is called the semiminor axis. A planet's mean distance is half the sum of the perihelion and aphelion distances. This is equal to the average distance of a planet from the Sun and also is the value of the semimajor axis. Kepler found that the cube of the mean distance for any planet is equal to the square of that planet's period. This equation is expressed mathematically as $p^2 = r^3$, where p is the planet's period in Earth years and r is the planet's mean distance from the Sun, expressed in terms relative to the Earth's mean distance, 150 million kilometers, or one astronomical unit (AU). If the Earth's mean distance equals 1.0 AU, then Mars's mean distance is 1.5 AU, Venus's mean distance is 0.72 AU, and so on. Planetary orbits obey this third Keplerian law of motion as a result of the central character of gravity as well as its inverse-square-law nature of gravity.

Newton later reformulated Kepler's three laws using more sophisticated mathematics than was available to Kepler. Newton's modification of the first law states that each planet has an elliptical orbit with the center of mass between it and the Sun at one focus. The "center of mass" is a point between the two bodies (the Sun and the orbiting planet) where their masses are essentially balanced. Mathematically, it is the point at which the product of mass times length is equal for the two bodies: $M_1 L_1 = M_2 L_2$, where M = mass, L = length from center of mass, and subscripts 1 and 2 referring to bodies 1 and 2. The Sun is such an extremely massive body that its center of mass with any planet lies near the Sun's own center. Therefore, the Sun does lie essentially at a focus of the planetary ellipse, as Kepler stated. Its movement around the center of mass (deep within its interior) is detectable only as a slight wobble. For bodies that are more comparable to one another in terms of mass, such as pairs of stars, these objects actually revolve around a common point that lies between them. Pluto and its similarly sized satellite Charon provide a good example of that effect. Because the masses of these bodies are similar, they revolve around a common point known as a the barycenter.

Newton revised Kepler's second law as follows: Angular momentum in a two-body system is constant when no net external torque is present. This law originally described the fact that planets move more rapidly when they are closer to the Sun compared to when they are farther away. Newton found that all bodies that rotate or move around some center have "angular momentum."

This quantity is expressed as a body's mass times its speed times its distance from the center of mass (mvr, where m = mass, v = linear speed, and r = distance from the center of mass). Because angular momentum is constant for any two-body system in the absence of a net external torque, if r becomes greater, v must become smaller to compensate (mass always remains constant). On the other hand, near the center of mass (the Sun, for planets), the distance r is diminished and speed v must increase to compensate. Conservation of angular momentum comes into play when a spinning skater pulls her outstretched arms close to her body, initiating a more rapid spin rate. Physicists and astronomers usually talk about planetary speeds of revolution or more properly angular velocity, which is the linear speed per unit distance from the focus. In such a discussion, angular momentum then involves the product of moment of inertia times angular velocity. There is no net torque acting on the planet revolving about the Sun, so this angular momentum expression is conserved or remains constant. That means that the distance from the Sun squared times angular velocity is an invariant throughout the planet's orbital motion.

Newton's revision of Kepler's third law is especially important. Newton discovered that the sum of the masses of the two bodies times the square of the period is proportional to the cube of the mean distance, which is expressed mathematically as $(M_1 + M_2)P^2 = a^3$. The masses must be expressed as a fraction of the Sun's mass for the calculation to be valid. The immediate consequence of this equation is that astronomers could now use this equation to calculate the masses of distant bodies given information on the mean distance and period of the orbiting bodies. In most instances, the mass of the smaller body (planet or satellite of a planet) may be neglected because that mass is so insignificant compared to the Sun's mass (1.99×10^{30} kilograms, or 332,943 times Earth's mass). Rearranging the equation gives $M_1 = a^3/p^2$. This equation can now be used to calculate the mass M_1 of any central body that has a satellite of mass M_2.

Applications

Consequences of Kepler's and Newton's laws of planetary motion and gravity had an impact on the scientific world not only during their own time but to this very day. The results of their work continue to be used by astronomers to solve problems. For example, the flight path of the Apollo astronauts to the Moon and back was calculated using all three of Kepler's laws. The energy required to propel the Saturn 5 rocket on its way and later to orbit the Moon was calculated using Newton's laws of gravity. The same can be said for all interplanetary spacecraft, such as Voyagers 1 and 2, which visited and photographed the outer planets, Jupiter, Saturn, Uranus, and Neptune. The two Voyager probes were assisted in their journeys by using the gravitational attraction of these massive planets to accelerate them toward their next target. Calculating gravity assists involves kinetic energy and gravitational relationships developed by Newton.

One of the most useful of Kepler's laws for planetary astronomers is the third law as modified by Newton. This law allows calculation of the mass of a massive body using data about mean distance and period of one or more of its satellites. It has been used to calculate the masses of all planets that have satellites (which excludes Mercury and Venus). One of the most difficult mass determinations was that for the dwarf planet Pluto and its satellite Charon. These bodies are so far away from Earth that Charon was discovered only in 1977. Its orbital characteristics were determined, with great difficulty, some time later. The similar masses of Charon and Pluto cause them to orbit a center of mass (a barycenter) that lies nearly halfway between them, but the location of that barycenter is somewhat closer to Pluto than it is to Charon. The third law was used to calculate both the mass of Pluto, using data from Charon's orbit, and the mass of Charon, using data for Pluto. These calculations show that both bodies have very low masses and are most likely composed of methane ice.

Another important consequence of the laws of planetary motion involves the survival of life on Earth. One theory suggests that periodic mass-extinction events—such as the demise of the dinosaurs—may have been caused by gigantic impacts of asteroids (rocky planetoids with diameters of less then 1,000 kilometers) or comets (asteroid-sized ice balls) with Earth. In the solar system, most asteroids are concentrated in a belt between Mars and Jupiter, whereas most comets originate in the outer regions of the solar system and beyond. Occasionally, collisions or gravitational perturbations from the massive gas giant planets, such as Jupiter or Saturn, cause asteroids and comets to assume orbits that carry them near Earth. All these bodies have sufficient kinetic energy to resist Earth's gravitational attraction, so that objects that graze Earth's orbit continue by without going into orbit around Earth. This fact explains why Earth and other relatively low-mass planets have few or no satellites (while the gas giants—Jupiter, Saturn, Uranus, and Neptune—have many). Therefore, the bodies that do strike Earth, causing extinctions and making huge craters if they strike land areas, must make a direct hit of a moving Earth. Chances of that occurring on a frequent basis fortunately are rather low, but not zero. Given the billions of years of the history of Earth, it is probable that an occasional body will crash into Earth with catastrophic consequences. The high kinetic energy of these bodies is converted into heat and shock waves upon impact, causing considerable destruction. Newton's laws of gravity and motion play a pivotal role, mostly in determining the trajectories of these dangerous visitors to the inner solar system. By the same token, Newton's laws reveal ways that gravity could be ingeniously used to push possible impacting bodies away from a trajectory that otherwise would have them intersect with Earth, thereby averting a possibly cataclysmic collision.

Context

The history of science closely parallels the development of astronomy in that the study of heavenly bodies and their relationship to Earth dominated philosophical and religious thinking for millennia. One of the first scientists to study religious thinking and astronomical phenomena seriously was the Greek philosopher Aristotle (384-322 B.C.E.). Unlike most of his contemporaries, Aristotle used some observations to prove his speculations. His major contribution to planetary motion studies was his belief that the natural state of matter is to seek the center of the Earth, which is why objects always fall when released above the Earth. Although erroneous, this and related ideas laid the groundwork for later studies by Galileo and Newton on the effects of gravity. Aristotle also believed, as did many others, that Earth was at the center of the universe. That the Sun and planets revolved around Earth in perfectly circular orbits was advocated first by his great mentor, Plato. Later, Aristarchus (c. 270 B.C.E.), a Greek astronomer, adopted the idea that the Sun is at the center of the known universe. That idea was forgotten until revived nearly two thousand years later by Nicolaus Copernicus, whose "heliocentric" model, published shortly after his death in a volume titled *De revolutionibus orbium coelestium* (1543; *On the Revolutions of the Heavenly Spheres*, 1952; better known as *De revolutionibus*),

describes a system in which the planets orbit the Sun in perfect circles. Although not completely accurate, the heliocentric model eventually supplanted the Earth-centered model of Aristotle and other philosophers.

In the middle of the second century C.E., Ptolemy (c. 100-178 C.E.), wrote a text called *Mathēmatikē syntaxis* (c. 150 C.E.; *Almagest*, 1948) in which he summarized all that was known about astronomy up to that time. This book influenced astronomical thinking for the next millennium. It included a model of the solar system that was quite accurate in predicting planetary positions. Using an idea first developed by Apollonius of Perga (c. 240-170 B.C.E.), Ptolemy declared that planets move in perfect circles around Earth. Nevertheless, planets moved simultaneously in smaller circles called "epicycles." These were necessary to explain why the outer planets occasionally seemed to reverse their normally eastern motion relative to the stars and move west. One of the great triumphs of Kepler's laws is that they provide an explanation for retrograde motion (Earth moves faster and overtakes the outer planets). Kepler, and before him Copernicus, laid the foundations for a scientific understanding of planetary motions that broke the hold on thinking imposed by the *Almagest*.

Sir Isaac Newton used ideas developed by Galileo and Kepler to quantify the knowledge of planetary motion and

---

Kepler's Laws of Planetary Motion

Johannes Kepler's three laws of motion, articulated in the first years of the seventeenth century, laid the foundation for Sir Isaac Newton's law of universal gravitation.

First Law: A planet orbits the Sun in an ellipse, with the Sun at one of the two foci.

Second Law: The line joining the planet to the Sun sweeps out equal areas in equal times as the planet travels around the ellipse.

Third Law: The ratio of the squares of the revolutionary periods for two planets is equal to the ratio of the cubes of their semimajor axes. That is, the time it takes a planet to complete its orbit is proportional to the cube of its average distance from the Sun. The farther from the Sun an object is, the more slowly it moves.

Johannes Kepler. (Library of Congress)

to explain these motions in terms of gravitational forces and kinetic energies. He published his ideas in *Philosophiae Naturalis Principia Mathematica* (1687; *The Mathematical Principles of Natural Philosophy*, 1729; best known as the *Principia*). Although it is now known that Newton's laws do not work well on atomic and subatomic scales (treated in the discipline of quantum mechanics) or in cases where bodies are moving relative to one another at very high speeds close to that of light (which later were addressed by Albert Einstein's relativity theory), Newton's laws work perfectly under everyday conditions on Earth and in the solar system.

Newton's and Kepler's time-tested laws continue to be used by astronomers and other space scientists to make predictions about planetary motions and interactions. Relativity is needed, however, to explain the advance of Mercury's perihelion as it orbits so close to the Sun. Spacecraft deep in the gravitational well of the Sun likewise require calculations involving relativity to maintain them on their proper courses.

John L. Berkley

Further Reading

Arny, Thomas T. *Explorations: An Introduction to Astronomy.* 3d ed. New York: McGraw-Hill, 2003. A general astronomy text for the nonscientist. Includes an interactive CD-ROM and is updated with a Web site.

Beatty, J. Kelly, Carolyn Collins Petersen, and Andrew Chaikin, eds. *The New Solar System.* 4th ed. Cambridge, Mass.: Sky, 1999. Filled with color diagrams and photographs, a popular work on solar-system astronomy and planetary exploration through the Mars Pathfinder and Galileo missions. Accessible to the astronomy enthusiast. Provokes excitement in the general reader, who gains an explanation of the need for greater understanding of the universe around us.

Consolmagno, Guy. *Worlds Apart: A Textbook in Planetary Sciences.* Englewood Cliffs, N.J.: Prentice Hall, 1994. A text accessible to college-level science majors and general readers alike. Presents explanations using low-level mathematics and also involves integral calculus where required. Demonstrates how the area of planetary science progresses by questioning previous understanding in the light of new observations.

Halliday, David, Robert Resnick, and Jearl Walker. *Fundamentals of Physics, Extended.* 9th ed. New York: Wiley, 2007. This textbook has taught millions of college students the fundamentals of physics. Its sections on Newton's laws of motion are particularly strong, as is the chapter on gravitation, which includes derivations of Kepler's laws of planetary motion. Even for those not familiar with basic calculus, there is much to be gained by studying from this all-encompassing work.

Hartmann, William K. *Moons and Planets.* 5th ed. Belmont, Calif.: Thomson Brooks/Cole, 2005. A college-level text that is clearly written; most nonspecialists should find this book a rich source of information. Chapter 3, "Celestial Mechanics," discusses the historical development and application of the laws of gravity and motion. Contains detailed black-and-white diagrams and photographs. Tables in the appendix offer comprehensive data on planetary orbital characteristics and other useful information.

Karttunen, H. P., et al., eds. *Fundamental Astronomy.* 5th ed. New York: Springer, 2007. A well-used university textbook in introductory astronomy. Contains some calculus-based treatments for those who find the standard treatise for typical ASTRO 101 classes too low level. Suitable for an audience with varied science and mathematical backgrounds. Covers all topics from solar-system objects to cosmology.

Leverington, David. *Babylon to Voyager and Beyond: A History of Planetary Astronomy.* New York: Cambridge University Press, 2003. Takes a historical approach to planetary science. Heavily illustrated, concluding with a summary of spacecraft discoveries. Suitable for general readers and the astronomy community alike.

McBride, Neil, and Iain Gilmour, eds. *An Introduction to the Solar System.* Cambridge, England: Cambridge University Press, 2004. A complete description of solar-system astronomy suitable for an introductory college course as well as nonscientists. Filled with supplemental learning aids and solved student exercises. A Web site is available for educator support.

Serway, Raymond A., Jerry S. Faughn, Chris Vullie, and Charles A. Bennet. *College Physics.* 7th ed. New York: Brooks/Cole, 2005. A textbook used at the introductory level in college physics courses, filled with sample problems, including those on laws of motion. Comes with an online teacher/student resource.

Snow, Theodore P. *The Dynamic Universe.* Rev. ed. St. Paul, Minn.: West, 1991. A general introductory text on astronomy. Covers the kinematics and dynamics of planetary motion. Features special inserts, guest editorials, and a list of additional readings at the end of each chapter. College level.

Stephenson, Bruce. *Kepler's Physical Astronomy.* Princeton, N.J.: Springer, 1994. A complete historical account of the search to understand the orbital behavior of planets in the solar system. For both the technical and general reader.

PLANETOLOGY: COMPARATIVE

Category: Planets and Planetology

Spacecraft have obtained detailed photographic, magnetic, radar, and chemical data from the planets Mercury, Venus, Mars, Jupiter, Saturn, Uranus, and Neptune as well as from numerous natural satellites, and even from some asteroids and comets. Data have been used in the preparation of models describing the structure and geological history of planetary and minor bodies throughout the solar system.

Overview

Comparative planetology is the study of the broad physical and chemical processes that operate in and on planets over time. It looks for patterns in the similarities and differences displayed by the planets and seeks to provide explanations for them in terms of planetary origins and evolution.

The first successful step in planetary exploration using robotic spacecraft was taken on August 26, 1962, when the National Aeronautics and Space Administration's (NASA's) Mariner 2 spacecraft was launched on a flyby mission to Venus. Mariner 4 was sent to Mars on November 28, 1964. Mariner, Pioneer, Pioneer Venus, Venera, Viking, Voyager, Magellan, Galileo, and Pathfinder space probes have sent back information about Mercury, Venus, Mars, Jupiter, Saturn, Uranus, and Neptune. Long after their primary missions had been completed, some of these spacecraft continued to transmit valuable information back to astronomers on Earth.

The planet Mercury had eluded detailed analysis by astronomers for centuries because of its small size and close proximity to the Sun. Mariner 10, launched in 1973 with a dual mission of studying the clouds of Venus and of photographing Mercury, made three passes by Mercury and was able to photograph about 45 percent of the surface of the planet. Mariner 10 was thus the first spacecraft to take scientific equipment to Mercury. Photographs of Mercury revealed a heavily cratered surface very much like that of Earth's moon. Naturally there are differences between the Moon and Mercury. Since the number of craters per square kilometer varies by as much as a factor of ten, it is believed that some craters may have been covered by a volcanic process. Still, Mercury's surface shows evidence of less volcanic activity than the Moon's. Mercury's largest impact basin, Caloris, has a diameter of 1,300 kilometers. It is believed to have been formed when a large asteroid struck the planet. Photographs show that

the shock wave from this collision penetrated the planet and altered the surface on the opposite side, an example of antipodal focusing of seismic energy by the planet's core. Compression scarps (cliffs), which can be as much as 3 kilometers tall and hundreds of kilometers long, were also found on Mercury. They are younger than the craters and are thought to have formed as a result of some internal process such as the cooling of the core of the planet. A magnetic field with a strength of about 1 percent of Earth's magnetic field and a very diffuse atmosphere containing mostly helium were found. Surface temperature ranges from 90 to 948 kelvins.

Venus has been a difficult planet to study from Earth because of its dense atmosphere. Russian Venera probes found a surface temperature of 748 kelvins and an atmospheric pressure of 95 atmospheres. (One atmosphere is the pressure exerted by Earth's atmosphere at sea level.) Photographs of the Venusian surface revealed some areas with smooth plains, while other areas have a rocky terrain. Radar mapping first by the American Pioneer Venus probes and later, in higher resolution and with fuller coverage, by the Magellan orbiter, shows a surface that is 70 percent gently rolling plains, 20 percent lowlands, and 10 percent highlands. The Ishtar Terra highland area of Venus is larger than the United States and includes the mountain Maxwell Montes, which stands about 11.3 kilometers tall. Alpha Regio and Beta Regio are mountainous regions that may contain shield volcanoes. A Russian Vega probe observed lightning discharges in these regions, which could mean that the volcanoes are still active. Lowland areas have the appearance of a cracked slab of lava or cemented volcanic ash. The rocks in the plains are probably granitic rock or potassium-rich basalt.

At the relatively low altitude of 26 kilometers, the Venusian atmosphere is clear, with the temperature dropping to 583 kelvins and the pressure to 20 atmospheres. A thick cloud layer, which is about 80 percent liquid sulfuric acid in the upper portion, exists from 26 kilometers to 60 kilometers above Venus's surface. A sulfuric acid haze exists from 60 kilometers to 80 kilometers altitude. The overall atmosphere is made up of 96 percent carbon dioxide, 3.4 percent nitrogen, and trace amounts of several other gases, and reflects 76 percent of the light striking it. No planetary magnetic field was found.

Following the Pioneer Venus probes after more than a decade's hiatus, NASA returned to Venus with the Magellan orbiter. The primary objective of Magellan was to obtain a high-resolution map of nearly the entire

The solar system planets, in a composite of images captured by missions from Mariner 10 to Cassini. (NASA/JPL)

Mars are concentrated in the southern hemisphere. The northern hemisphere, which has been smoothed by lava flows, has fewer craters, and their features tend to be sharper, indicating that they may be younger than those found in the southern hemisphere. Many of the craters on Mars show evidence of significant erosion. Mars Exploration Rover studies of rocks in situ revealed evidence of sedimentary processes requiring the presence of water.

Mars's seasonal polar caps, made of carbon dioxide, extend well down into the hemisphere, experiencing winter, but shrink and retreat quickly in early summer. Residual polar caps remain throughout the year, although their size does vary. The southern cap is made of carbon dioxide only. The northern polar cap is larger than the southern cap, has a wider temperature variation, and contains mostly water ice. It may be one of the main storehouses for water on Mars. The Mars Polar Lander was designed to investigate that, but it crashed. In 2008, the Mars Phoenix landed on the northern polar region to continue that search for water; by sampling and analyzing the subsurface material, it provided direct evidence that water ice was present in significant amounts.

Mars's atmosphere is 95.3 percent carbon dioxide, 2.7 percent nitrogen, and 1.6 percent argon, with a total pressure of 0.01 atmosphere. The atmospheric temperature varies from 243 down to 173 kelvins. Sublimation of the polar ice caps causes the pressure to vary about 20 percent from season to season. Fog forms in low areas in the early morning. Clouds have been seen around some of the volcanoes. Winds with speeds of at least 150 kilometers per hour pick up surface dust and cause global dust storms. It can take as long as six months for all the dust to settle out from one of these storms. Several spacecraft in orbit at Mars have had to wait for months until the dust cleared in order for their cameras to resume imaging surface features. Long-lived orbital spacecraft have taken images of certain features at widely spaced intervals, showing evidence of wind erosion having altered the surface. Other

globe of Venus using a synthetic aperture radar system. Magellan's mission was extended to permit more site-specific investigations. In all, Magellan produced a map of Venus's surface that in resolution and coverage exceeded any available maps of Earth's surface at that time.

Spacecraft such as Mariner, Viking, Pathfinder, and the Mars Exploration Rovers Spirit and Opportunity found that the surface of Mars contains craters, large plains marked by great sand dune areas, chaotic terrain characterized by irregular ridges and depressions, and many volcanoes. The largest volcano, Olympus Mons, stands 27 kilometers tall, has a base diameter of 600 kilometers, and is 64 kilometers across its summit. Orbiting probes uncovered unmistakable signs of catastrophic floods. Impact craters on features have shown evidence of water slumping of crater walls in recent times; that interpretation remains under consideration, however.

The atmosphere of Jupiter is thought to be about 1,000 kilometers thick, with a gaseous composition of 75 percent hydrogen, 24 percent helium, and 1 percent other gases. Pressure at the base of such an atmosphere would be about 100 atmospheres, and the temperature would be about 813 kelvins. The temperature at the top of the atmosphere is only about 113 kelvins. Colored bands, termed zones and belts, are visible in the atmosphere. Zones are yellow-white and represent high-pressure areas where warm currents are rising. Belts are brown, red, or blue-green and represent low-pressure areas where colder gases are sinking. Colors have their source in the interaction of chemical compounds in the atmosphere with sunlight. Very strong wind currents flow in opposite directions where the belts and zones touch. Bands are stable and have not changed their positions for the past one hundred years.

Jupiter's Great Red Spot has been its most prominent feature for over 350 years. It is about 26,000 kilometers from east to west and 14,000 kilometers from north to south. This large cyclonic storm wanders in an east-west fashion and may be stable enough to last for many more centuries. Voyager investigations were followed by the Galileo orbiter

and its atmospheric probe, which entered the atmosphere at a point where it found relatively little water vapor.

Jupiter's magnetic field, which is at least ten times stronger than Earth's, produces a radiation belt that is strong enough to kill a human quickly. The radiation belt almost ruined some transistors in the Pioneer probes. Jupiter's radio emissions come from charged particles trapped in the magnetic field. Voyager 1 found a ring system whose main ring is 6,000 kilometers wide and 30 kilometers thick. A thin sheet of material extends to the surface of the planet.

Jupiter has sixty known moons. Fourteen were discovered by Earth-based astronomers, while two were found in Voyager 1 photographs. Others were found by Voyager 2, the Hubble Space Telescope, and Galileo spacecraft. Active volcanoes were found on the moon Io, Jupiter's innermost moon. An icy crust on Europa is believed to cover an ocean of liquid water; evidence of crustal movement upon such a water layer was found in 2008. That provided strong evidence for internal heating to drive large-scale movements of the icy crust.

Saturn is the second of the giant planets visited by spacecraft. Its atmospheric structure is much like that proposed for Jupiter, but its composition is more like the Sun's, with only 11 percent helium. Belts and zones seen on Jupiter are also visible on Saturn, but their colors are

Comparative Data on the Planets of the Solar System

Parameter	Mercury	Venus	**Earth**	Mars	Jupiter
Mass (10^{24} kg)	0.3302	4.8685	**5.9742**	0.64185	1,898.6
Volume (10^{10} km^3)	6.083	92.843	**108.321**	16.318	143,128
Equatorial radius (km)	2,439.72	6,051.8	**6,378.1**	3,396	71,492
Ellipticity (oblateness)	0.0000	0.000	**0.00335**	0.00648	0.06487
Mean density (kg/m^3)	5,427	5,243	**5,515**	3,933	1,326
Surface gravity (m/s^2)	3.70	8.87	**9.80**	3.71	24.79
Surface temperature (Celsius)	−170 to +390	+450 to +480	**−88 to +48**	−128 to +24	−140
Satellites	0	0	**1**	2	14
Mean distance from Sun millions of km (miles)	58 (36)	108 (67)	**150 (93)**	228 (141)	779 (483)
Rotational period (hrs)*	1,407.6	−5,832.5	**23.93**	24.63	9.9250
Orbital period	88 days	224.7 days	**365.25 days**	687 days	11.86 yrs

not as intense. The outer layer is predominantly hydrogen. The temperature at the bottom of this layer is 70 kelvins.

Saturn has the most highly developed ring system in the solar system. The ring system has a width of 153,000 kilometers and a thickness of 2 kilometers. There are nine distinct rings, labeled A through G. (Identification of portions of the planet's ring system retains naming schemes from the early days of telescopic observations, before the full complexity of the rings was seen; as a result, six parts of the overall ring system are identified by capital letters, whereas the rest are given names. Unfortunately, letters and names do not necessarily provide information as to distance from the outer atmosphere; for example, the C and B rings are outside the D ring but inside the A ring, and the F and G rings are outside the A ring.) The rings are very complex, made up of an extremely large number of ringlets, some of which are only about two kilometers wide. Shepherding moons orbit around the edge of some rings, and their gravity functions to maintain the sharp edge on the rings. The B ring shows dark features that resemble spokes in a wheel. The particle size in the rings varies from a few thousandths of a centimeter to about ten meters. The spokes rotate as if solid and appear to be particles electrostatically raised above the ring plane.

Saturn has at least sixty satellites. The Cassini spacecraft provided images of many of those satellites detected after the Voyager era. Saturn's only large satellite, Titan,

has received a great deal of attention, since it is able to retain a thick atmosphere of nitrogen, methane, and other hydrocarbons. Its surface is obscured due to the thickness of that atmosphere. For that reason, the Cassini orbiter carried a European Space Agency probe named Huygens that was detached from the main spacecraft in order to land on the surface of Titan. Huygens provided evidence of liquid hydrocarbons at its touchdown site existing at cryogenic temperatures. Cassini then was able to image ancient shorelines and prove the existence of lakes of liquid methane across the surface of Titan.

In January, 1986, Voyager 2 passed by Uranus en route to Neptune. Uranus's rotational axis is tilted 82° from the plane in which it orbits. Its rotational direction is retrograde. Voyager 2 data established the rotational period of the planet to be 17 hours, 14 minutes. The greenish atmosphere of Uranus is unusually free of clouds. The primary components of the atmosphere are hydrogen (84 percent), helium (14 percent), and methane (2 percent). The temperature of the atmosphere where the pressure is 1 atmosphere is about 73 kelvins. Voyager 2 observed a tenuous haze around Uranus's rotational pole. This haze is probably formed by the steady irradiation of the planet's upper atmosphere by solar ultraviolet light. Uranus is a weak emitter of thermal radiation from deep within the atmosphere. The planet has been found to be warmer than thought, which implies a greater transparency of the

Comparative Data on the Planets of the Solar System *(continued)*

Parameter	Saturn	Uranus	Neptune	Pluto (dwarf planet)
Mass (10^{24} kg)	568.46	86.832	1,102.43	0.0125
Volume (10^{10} km³)	82.713	6,833	6,254	0.715
Equatorial radius (km)	60,268	25,559	24,764	1,195
Ellipticity (oblateness)	0.09796	0.02293	0.01708	0.0000
Mean density (kg/m³)	687	1,270	1,638	1,750
Surface gravity (m/s²)	8.96	8.69	11.00	0.58
Surface temperature (Celsius)	−160	−180	−200	−238
Satellites	11	5	2	1
Mean distance from Sun millions of km (miles)	1,426 (884)	2,869 (1,779)	4,494 (2,786)	5,899 (3,657)
Rotational period (hrs)*	10.656	−17.24	16.11	−153.3
Orbital period	29.46 yrs	84.01 yrs	164.80 yrs	247.70 yrs

atmosphere than models had predicted. The magnetic field of Uranus is inclined at a 60° angle to the axis of rotation. (Earth's rotational axis and magnetic field are roughly parallel by comparison.)

Voyager 2 found a large spot on Neptune's southern hemisphere in 1989 similar to Jupiter's Great Red Spot. The probe also confirmed the presence of three thin, faint rings around the planet and a magnetosphere. The atmosphere is cold, about 53 kelvins, and its soft, blue tint comes from the presence of methane in the upper atmosphere. Neptune has thirteen known moons, the largest of which, Triton, is covered with methane and nitrogen ices.

The New Horizons spacecraft was launched in January, 2006, to fly by the Pluto-Charon system and thereby complete the initial reconnaissance of all major systems of the solar system. New Horizons was launched when Pluto was still classified as a planet. Later that same year, a new identification system adopted by the International Astronomical Union (IAU) demoted Pluto to the status of a dwarf planet. In June, 2008, the IAU again redefined Pluto, this time as a plutoid, or plutino. Regardless of whether Pluto is a full-fledged planet or a plutoid, New Horizons will provide the first in-depth investigations and closeup images of Pluto and its nearly similar sized satellite Charon sometime in the second decade of the twenty-first century.

Knowledge Gained

Spacecraft data concerning the atmospheric composition and structure of individual planets have provided significant insight into the solar system as a whole. Mercury's small size and high temperature made it an unlikely candidate for having any measurable atmosphere, yet Mariner 10 found a tenuous atmosphere on the planet. This condition probably arises from the solar wind that bathes Mercury. Venus has a high surface temperature but significantly more mass than Mercury and has been able to retain its atmosphere effectively. Venus's high temperature prevents the buildup of any significant quantity of water, so that carbon dioxide remains in the atmosphere rather than forming carbonates as it can on water-rich Earth. Mars perhaps once had a much denser atmosphere, with large quantities of liquid water—possibly enough to cover the planet to an average depth of 10 meters. Channels on the surface point to large amounts of flowing liquid. As a result of low temperature and low surface gravity, most of Mars's atmosphere has been lost. Perhaps water ice is still trapped below the surface or in the north polar ice cap. Confirming that was the primary objective of the Mars Phoenix mission in 2008, and early results from the

lander strongly suggested that white material just underneath the soil was indeed water ice and neither salts nor dry ice. Mars Phoenix was outfitted with a Thermal Evolution and Gas Analyzer (TEGA). Before the end of 2008, TEGA obtained evidence of the presence of water vapor after heating soil samples that were carefully placed within its ovens by a robotic arm equipped with a scoop.

Since the giant planets Jupiter, Saturn, and Uranus, and Neptune have much more massive cores and are much colder than the four inner planets, they can retain light gases such as hydrogen and helium effectively. Differences exist among these four, however, because the core size differs from planet to planet.

The terrestrial planets, Mercury, Venus, Earth, and Mars, show many similar surface features—craters, volcanoes, and mountains. Only Earth has shown activity of its volcanoes, but discharges of lightning around the volcanic mountains on Venus suggests that they may be active also. Volcanism has also been found on Jupiter's satellite Io, Saturn's satellite Enceladus, and Neptune's satellite Triton.

The giant planets all have ring systems, although each system is different from the other three. Saturn's rings are extremely complex, with small divisions between the rings. Uranus has a set of narrow ribbons separated by large spaces, while Neptune has only partially complete ring arcs. Jupiter has a three-component ring system. The innermost portion is called the Halo Ring. Further out is the Main Ring, and that is followed by the wispy Gossamer Rings. High-resolution images from Galileo at Jupiter, Cassini at Saturn, and Voyager 2 at Uranus and Neptune greatly added to the storehouse of knowledge about diversity in ring system dynamics.

One great hope in the exploration of Mars was that some life-form would be discovered. Experiments performed by the Viking landers provided no definitive results. Many astronomers believe that Mars's environment is much too harsh presently to support life as it would exist on Earth. Any primitive non-Earth-like forms of life might be difficult to detect. Life, primitive or otherwise, may also be possible on Europa, Enceladus, or Titan. Few scientists expect to find organisms on the latter two satellites, but some hold out hope that some degree of organized life-forms may be swimming in Europa's ocean under the satellite's icy crust. Until a Europa lander equipped with a subterranean probe can be sent to this satellite, however, that remains only wishful speculation on the part of exobiologists.

Context

Fascination with outer space is evident when one examines the popularity of space-based science-fiction books, films, and television programs, and when one keeps track of the number of Internet hits on NASA Web sites during high-profile missions like Mars Exploration Rover landings on the Red Planet or space shuttle flights to refurbish the Hubble Space Telescope. Solar system exploration programs are scientific attempts to satisfy human curiosity about space. One fundamental purpose for planetary exploration is to seek a better understanding of the history and perhaps the origin of the solar system. While current models meet some of the criteria, many questions remain. Examination of planetary atmospheres, magnetic fields, ring systems, satellites, and surfaces allows models to be improved and planetary history to be more accurately recorded. For example, the Jupiter and Saturn systems are large enough for them and their satellites to constitute small-scale solar systems. Study of such smaller systems could reveal significant details about the solar system as a whole.

Humanity also has a desire to know whether life exists in any place other than Earth. Are we alone? Is the vastness of the universe just for us, or is it teeming with life? Chances of detecting life in another star system are remote, even if it does exist. The search on the planets of the solar system is much more easily accomplished. In the late twentieth century, both the United States and the former Soviet Union planned uncrewed missions to Mars that would include orbiters, landers, balloons, surface-roving vehicles, and a round-trip mission to return soil samples to Earth. Many of those ambitious plans were delayed considerably, but early in the new millennium an armada of robotic spacecraft orbited around the Red Planet and a number of landers were on the surface searching for evidence of water.

Dennis R. Flentge

Further Reading

Bagenal, Fran, Timothy E. Dowling, and William B. McKinnon, eds. *Jupiter: The Planet, Satellites, and Magnetosphere.* Cambridge, England: Cambridge University Press, 2004. A comprehensive work about the biggest planet in the solar system, comprising a series of articles by experts in theirs field of study. Excellent repository of photography, diagrams, and figures about the Jupiter system and the various spacecraft missions that have unveiled its secrets.

Beattie, Donald A. *Taking Science to the Moon: Lunar Experiments and the Apollo Program.* Baltimore: Johns Hopkins University Press, 2003. Explains the science gleaned from the Apollo lunar landings, including the Apollo Lunar Surface Science Experiment Packages (ALSEPs) and their results.

Briggs, G. A., and F. W. Taylor. *The Cambridge Photographic Atlas of the Planets.* New York: Cambridge University Press, 1982. A collection of the best photographs taken by space probes from the United States and the Soviet Union. In addition to the captions accompanying the photos, a discussion of the important features of each planet and its satellites is provided.

Greenberg, Richard. *Europa the Ocean Moon: Search for an Alien Biosphere.* New York: Springer, 2005. A complete description of current knowledge of Europa through the post-Galileo spacecraft era. Discusses the astrobiological implications of an ocean underneath Europa's icy crust. Well illustrated and readable by both astronomy enthusiasts and college students.

Grinspoon, David Harry. *Venus Revealed: A New Look Below the Clouds of Our Mysterious Twin Planet.* New York: Basic Books, 1998. A thorough examination of the geology of Venus that incorporates Magellan mapping and other data. Explains the Venusian greenhouse effect. A must for the planetary science enthusiast who wants an integrated approach to science and history. Includes speculation about Venus's past.

Harland, David M. *Cassini at Saturn: Huygens Results.* New York: Springer, 2007. Essentially a complete collection of NASA releases from the start of Cassini flight operations through the majority of Cassini's seventy orbits of its primary mission. Provides a thorough explanation of the entire Cassini program, including the Huygens probe's landing on Saturn's largest satellite. Cassini's primary mission concluded a year after this book was published. Technical but accessible to a wide audience.

_____. *Water and the Search for Life on Mars.* New York: Springer Praxis, 2005. A historical review of telescope and spacecraft observations of the Red Planet up through the Spirit and Opportunity rovers. Covers all aspects of Mars exploration but focuses on the search for water, believed to be the most necessary ingredient for life.

Hartmann, William K. *Moons and Planets.* 5th ed. Belmont, Calif.: Thomson Brooks/Cole, 2005. An updated version of a classic text that covers all aspects of planetary science. Particularly strong in its presentation of Earth-Moon science. Takes a comparative planetology approach rather than providing individual chapters on each planet. Examines atmospheres, magnetospheres, satellites, and interiors of the solar system's planets.

Irwin, Patrick G. J. *Giant Planets of Our Solar System: An Introduction.* 2d ed. New York: Springer, 2006. Suitable as a textbook for upper-level college courses in planetary science. Focuses on Jupiter, Saturn, Uranus, and Neptune and their satellites, rings, and magnetic fields. Filled with figures and photographs.

Lovett, Laura, Joan Harvath, and Jeff Cuzzi. *Saturn: A New View*. New York: Harry N. Abrams, 2006. A coffee-table book with about 150 of the best images returned by the Cassini mission to Saturn. Covers the planet, its many satellites, and the complex ring systems.

Morrison, David, and Tobias Owen. *The Planetary System*. 3d ed. San Francisco: Pearson/Addison-Wesley, 2003. A discussion of data from each of the planets accompanied by a large number of photographs and line drawings. Although intended as a college-level astronomy textbook, it provides good reading for anyone with an interest in the solar system.

Squyres, Steve. *Roving Mars: Spirit, Opportunity, and the Exploration of the Red Planet*. New York: Hyperion, 2006. Written by the principal investigator for the Mars Exploration Rovers Spirit and Opportunity, this fascinating book provides a general audience with a behind-the-scenes look at how robotic missions to the planets are planned, funded, developed, and flown. A personal story of excitement, frustrations, a scientist's life during a mission, the satisfaction of overcoming difficulties, and the ongoing thrills of discovery.

PLANETOLOGY: VENUS, EARTH, AND MARS

Category: Planets and Planetology

The rocky planets Venus, Earth, and Mars are similar in size, mass, and proximity to the Sun, yet they have evolved in three very different directions. Venus has a thick atmosphere of carbon dioxide and a surface temperature hot enough to melt lead. Earth's atmosphere is mostly nitrogen with only a trace amount of carbon dioxide. Mars has a very thin atmosphere of carbon dioxide and quite cold surface temperatures.

Overview

When the planets formed in the primordial solar system, they existed primarily as gases coalesced by gravity. The four inner planets, because of their proximity to the Sun, lost most of their primitive atmosphere of hydrogen and helium, retaining only a molten rocky core rich in heavier elements. As these planets cooled, a secondary atmosphere was created by gases ejected from the mantle by many active volcanoes on the geologically unstable surface. These gases included ammonia, carbon dioxide, water vapor, and nitrogen. Mercury, being the smallest planet with the greatest intensity of sunlight, quickly lost its entire atmosphere.

More massive and farther from the Sun, Venus retained its secondary atmosphere. This caused the surface to remain hot enough to prevent the water vapor from condensing into liquid. As the concentration of CO_2 increased, the heat in conjunction with ultraviolet light from the Sun dissociated the water vapor into free hydrogen and oxygen. The hydrogen dissipated into space, while the very reactive oxygen combined with rock minerals, thus disappearing from the atmosphere. Estimates made from Venus's rocks indicate that initially there was enough water vapor on Venus to have covered the surface with an ocean at least 9.14 meters deep, on average. The continued volcanic outgassing of CO_2 continued to heat the surface until equilibrium was established. When visible light strikes a planetary surface, much of the energy is converted into heat, which radiates from the surface as infrared radiation. Carbon dioxide will absorb outgoing infrared radiation and reemit it uniformly in all directions, thus heating the surface in proportion to the amount of CO_2 present in the atmosphere. Because Venus's atmosphere is 95 percent carbon dioxide, the surface temperature is even hotter than the surface of the planet Mercury, due to what is referred to as a "runaway greenhouse effect." Venus therefore is now a stifling inferno smothered by a thick carbon dioxide (CO_2) atmosphere. The surface temperature is about 750 kelvins, distributed uniformly across the planet due to atmospheric carbon dioxide.

Because Earth is 43 percent farther from the Sun, the solar intensity is only half that at Venus, while the gravitational attraction is slightly greater than Venus's. Consequently, water vapor was retained and able to condense as the planet cooled, thereby forming the oceans. Carbon dioxide readily dissolves in warm water. It was progressively removed from the atmosphere, eventually forming carbonate rocks. Relatively early in Earth's history primitive life formed in the oceans, and evolved into small shelled sea creatures. These primitive animals formed their shells from the dissolved carbon dioxide and calcium, thus further removing CO_2 from the water and leaving shell deposits which eventually agglomerated into calcium carbonate ($CaCO_3$). As CO_2 was removed from the water, more could enter from the atmosphere until almost all was gone. Analysis shows that the carbonate rocks on Earth's crust contain about the same amount of CO_2 as Venus's atmosphere. Although oxygen was present in Earth's primitive atmosphere, most reacted with metals, such as iron, to form oxides. Life further altered the atmosphere with photosynthesis commencing about 2.5 billion years ago, eventually reducing the concentration of atmospheric CO_2 to its present 0.04 percent and creating most of the free oxygen.

Mars is 1.5 times farther from the Sun than Earth. Being considerably smaller and less massive, Mars had less internal heating and consequently considerably less outgassing. Mars therefore has an extremely thin atmosphere, 95 percent of which is CO_2. This is an amount insufficient to create any significant greenhouse warming.

The Viking spacecraft showed that although Mars originally had a denser atmosphere and running water on the surface, most of the atmosphere leaked off into space because of the low gravity. The atmospheric pressure, less than 1 percent of Earth's sea-level pressure, is below the pressure where water can exist in the liquid state. Although a core of water ice is present in the polar caps, the ice in those caps consists primarily of solid carbon dioxide. Surface temperatures on Mars vary from a maximum of 300 kelvins at the warmest spot at the warmest moment of the warmest day of the Martian year to typical nighttime temperatures of 155 kelvins. At a location midway between the equator and the poles, the maximum daytime temperature barely exceeds the freezing point of water. Much of the remaining water is assumed to be tied up below the surface as permafrost. The original CO_2 is now either carbonate rocks, or dissipated into space. The polar caps consist of small permanent water ice caps and solid CO_2 (dry ice), which sublimates directly into its vapor form when the temperature increases during summer.

Although the atmospheres of Venus, Earth and Mars appear very different, when the elements bound up in rocks and permafrost are included, the inventories of water, carbon dioxide, and nitrogen are remarkably similar when adjusted for the differing planetary masses.

The topography of the three planets is a result of their size and evolution. Earth, the largest, is dominated by rolling seafloor plains interrupted by continents and mountain ranges where continental plates have collided. All traces of Earth's primeval crust have been destroyed by basalts that erupted to form much of the seafloor crust and by plate tectonics that broke up and recycled the original surface.

On Venus, basaltic volcanism dominated and covered most of the planet with lava flows. Only a small part of the primordial surface remained as protocontinents projecting several miles above the plains. Perhaps because of its smaller size, there was not enough internal energy to drive plate tectonic crustal motions. Some volcanoes may still be active. The scarcity of meteorite impact craters indicates that the entire crust has been replaced within the past half million years—recently, in geologic terms.

Being considerably smaller than Venus, Mars lacked sufficient tectonic energy to destroy its original cratered features. The deepest surface depressions are ancient impact craters but the highest Martian mountains are simply masses of volcanic lava surmounted by the volcanic caldera. As a rule of thumb, small worlds preserve their ancient surfaces formed by meteorite bombardment, but volcanic forces break through and resurface parts of the planet. Larger planets, on the other hand, have surfaces dominated by the internal forces of volcanism and plate tectonics.

Knowledge Gained

It has been known since the eighteenth century that Venus has an atmosphere. Although it was first assumed to be composed of water vapor, in 1932 spectroscopic studies indicated that Venus's atmosphere was primarily CO_2. Thermal radiation measurements in the 1960's indicated a surface temperature close to 750 kelvins. This temperature was confirmed when, in 1967, the Russian probe Venera 4 crashed into the Venusian surface. In 1970 the Venera 7 successfully landed on Venus's surface and transmitted data for twenty-three minutes, verifying the surface temperature and an extremely dense CO_2 atmosphere. The American spacecraft Pioneer (1978) discovered clouds of sulfuric acid droplets positioned about 48 kilometers above surface. This mission also included an orbiter that mapped the surface features by using radar. Russian landings later sent back panoramic photographs of a haze-free surface imbued with a reddish hue from atmospheric filtered sunlight, strewn with boulders on gravel and fine, rocky soil. In 1985, two balloons dropped by Russian probes floated in the sulfuric acid clouds for forty-six hours, measuring hurricane-force winds of 240 kilometers per hour but temperatures and pressures similar to Earth's surface. When the sulfuric acid particulates reach a sufficient size, they begin to fall out of the cloud deck as sulfuric acid rain. Unlike rain on Earth, however, they never reach the ground. The rapidly increasing temperature cause them to evaporate.

Telescopic observations of Mars during the eighteenth and nineteenth centuries revealed clear seasonal changes.

In summer the polar cap shrinks, while dark markings, once assumed to be vegetation, darken and grow more prominent. This assumption was disproved in 1965 when Mariner 4 returned the first close-up pictures revealing a desolate, red-colored desert. Three Russian probes reached the surface between 1971 and 1974, but all failed to return useful data. The first successful landing was by Viking 1 on July 20, 1976, followed in September by its sister craft, Viking 2. The Viking landers photographed a surface strewn with boulder-sized fragments of lava flows and meteorite impact craters in different stages of erosion. Soil analyses provided chemical evidence that Mars's atmosphere was once almost as dense as Earth's, which would have enabled water to flow. Earlier, Mariner 9 had photographed channels looking like dry riverbeds containing tributaries and sedimentary deposits, providing additional evidence that water once flowed on Mars.

Indisputable evidence of Mars's water was provided during the summer of 2008, when National Aeronautics and Space Administration's (NASA's) Mars Phoenix Lander dug into Martian soil and scooped up what appeared to be ice. When melted it proved to be water vapor. Three spacecraft landed between 1997 and 2004 to photograph and analyze the surface using robotic rovers. The Martian surface consists of two parts: a heavily cratered southern highlands and northern smooth lowlands. Evidence from the Odyssey spacecraft (2001) suggests these lowlands were once filled by an ocean of liquid water, now present in permanent ice caps and subterranean permafrost. Mars's early dense atmosphere may have been depleted by changes in the polar climate. If solar radiation striking the poles were 10 percent higher in the past, the ice caps would have been much smaller and most of the now frozen CO_2 would have existed in a gaseous state.

Context

Although Venus, Earth, and Mars are similar in mass and distance from the Sun, enough minor differences existed to compel each to have evolved along very different paths. The atmospheres of Venus and Mars have been thoroughly analyzed, and their surface features have been carefully mapped, photographed, and subjected to chemical analysis. Nevertheless, several questions remain unanswered. How much water was initially present on Venus, and what happened to it? Why was there not more plate tectonic activity during Venus's formative years? Is Venus still volcanically active? Why does Earth's atmosphere contain an abundance of nitrogen, and why is it scarce on the other planets? Why did Mars have a much warmer surface in the past, and where is the once prevalent water? Did the formation of the gigantic volcano, Olympus Mons, cause a change in the tilt of the Martian axis that instigated a planetary cooling? Is Mars still volcanically active, or has it cooled sufficiently that new volcanoes cannot form?

Obviously, there is still much to learn about the comparative planetology and evolution of these three planets. Future interplanetary spacecraft that land on Venus and Mars will continue to advance the understanding of the dynamics of their surfaces and, in particular, the manner in which subterranean features influenced their respective planetary evolution.

George R. Plitnik

Further Reading

Chaisson, Eric, and Steve McMillan. *Astronomy Today*. 6th ed. New York: Addison-Wesley, 2008. Includes an excellent summary of the latest knowledge about the terrestrial planets, particularly Venus, Earth, and Mars. The atmosphere, the surface, and the internal structure of each of these planets is discussed in great detail, rendering the comparative planetology a straightforward exercise for the dedicated reader.

De Pater, Imke, and Jack J. Lissauer. *Planetary Sciences*. New York: Cambridge University Press, 2001. A challenging and thorough text for students of planetary geology. Covers extrasolar planets and provides an in-depth contemporary explanation of solar-system formation and evolution. Best for the most serious reader with a strong science background.

Encrenaz, Thérèse, et al. *The Solar System*. New York: Springer, 2004. A thorough exploration of the solar system from early telescopic observations through the space missions that had investigated all planets with the exception of Pluto by the publication date. Takes an astrophysical approach to place our solar system in a wider context as just one member of similar systems throughout the universe.

Esposito, Larry W., Ellen R. Stofan, and Thomas E. Cravens, eds. *Exploring Venus as a Terrestrial Planet*. New York: American Geophysical Union, 2007. This is a collection of articles covering all major areas of Venus research. Technical.

Hartmann, William K. *Moons and Planets*. 5th ed. Belmont, Calif.: Thomson Brooks/Cole, 2005. This authoritative and regularly updated text considers all the major planetary objects in the solar system. The material is presented by grouping objects under unifying principles, thus elucidating their similarities and their differences as well as the physical processes behind their evolution. Although most of the material is descriptive, some algebra and elementary calculus are included.

Lewis, John S. *Physics and Chemistry of the Solar System.* 2d ed. San Diego, Calif.: Academic Press, 2004. Suitable for an undergraduate course in planetary atmospheres, but accessible to the general reader with a technical background.

McSween, Harry. *Stardust to Planets.* New York: St. Martin's Griffin, 1993. A well-written book that imparts geologic and atmospheric facts about the planets of the solar system in an easy-to-grasp manner. Venus and Mars are covered in considerable detail, and the information about Earth's geology makes planetary comparisons relatively straightforward.

Sagan, Carl. *Cosmos.* New York: Wings Books, 1995. Based on the television series of the same name, this lavishly illustrated classic includes considerable information about the evolution of surface features and atmospheres of Venus and Mars.

Seeds, Michael A. *Foundations of Astronomy.* 9th ed. Belmont, Calif.: Thomson Brooks/Cole, 2007. This lavishly illustrated text commingles experimental evidence and theory to provide deep, but well-explained, elucidations of many fascinating aspects of the universe. In particular, the five chapters on comparative planetology contain a plethora of data, as well as beautiful pictures, from various spacecraft encounters with these planets since the 1970's.

VAN ALLEN RADIATION BELTS

Category: Earth

The Van Allen radiation belts are concentrated rings of ionized particles in Earth's magnetosphere. Detailed study of the radiation belts led to an understanding of certain phenomena occurring in the ionosphere and the determination of the physical properties of the exosphere.

Overview

The Van Allen radiation belts are concentrated, torus-shaped regions of charged particles within Earth's magnetosphere. These particles, made up of protons, electrons,

and other ions, spiral about in great numbers between Earth's magnetic poles. The magnetic and charged particles within the Van Allen belts can be divided into four regions: the Van Allen geomagnetically trapped radiation region, the auroral region, the magnetosheath, and interplanetary space. The inner and outer belts are part of the Van Allen geomagnetically trapped radiation region. In discussions of the Van Allen belts, the magnetic storm is often referred to as a third radiation belt.

The inner zone stretches from about 1,000 to more than 5,000 kilometers above Earth. It is mainly independent of time. Its composition is nearly consistent with that expected for the decay products of cosmic-ray-produced neutrons in the atmosphere (a neutron is an elementary, neutral particle of mass); this zone is of cosmic-ray origin. The radiation in the middle of the inner zone is composed of electrons with energies exceeding 40 kilo-electron volts (keV) and protons with energies greater than 40 million electron volts (MeV). (Electrons are elementary particles with a negative charge; protons are positively charged elementary particles.) In the inner belt, many of the high-energy protons are capable of penetrating several inches of lead. At the edge of the inner zone, in the region of geomagnetic latitudes 35° to 40°, low-energy electrons are found. The decay of light-scattering neutrons gives rise to high-energy protons. Beyond Earth's magnetic field, the mean ionizing capacity is 2.5 times higher than the minimum ionizing capacity. Particles in the inner zone are stable and exist for a long period of time.

The outer zone stretches about 15,000 to 25,000 kilometers above Earth. This zone undergoes very large temporal fluctuations appearing to be caused by solar activity and auroras, atmospheric heating, and magnetic storms. The outer belt contains soft particles; it is of solar origin. The outer zone contains electrons with more than 40 keV in energy and protons with more than 60 MeV in energy. The outer zone has greater geophysical significance than the inner zone. According to the comparison of E. V. Gorchakov, the boundaries of the outer zone coincide with isochasms (lines of equal probability of auroras). Trapped particles introduce magnetic effects in the outer radiation belt. This effect was measured by Luna 1. The increase in ionization of the outer zone is unstable. Particles exist for a short period of time compared with those of the inner belt.

A third radiation belt is produced by magnetic storms. Protons are transported from the Sun in a corpuscular stream and injected by magnetic field perturbations into Earth's field. The charge exchange with neutral hydrogen in Earth's exosphere is the fastest mechanism of removal. This is about a hundred times faster than scattering from ions in the exosphere. With the exception of trapped radiation, the entire region in the magnetic cavity is known as the auroral region. Auroral particles, the islands or pulses in the long tail and spikes at high latitudes of 1,000 kilometers, are phenomena that occur in the auroral region. Electrons of uniform angular distribution have a roughly constant intensity between 100 and 180 kilometers in altitude.

The magnetosheath lies between the shock front formed by the solar wind and the magnetic cavity. Islands of electrons have been observed in the magnetosheath. At its widest, the magnetosheath is about four times the radius of Earth. It contains a compressed, seemingly chaotic interplanetary magnetic field. The interplanetary field connected to the Sun is predominantly in the ecliptic plane. The field terminates when the solar wind undergoes a shock transition to subsonic flow.

The lifetime of trapped particles decreases with distance from Earth. The lifetime of electrons with energies greater than 1 MeV at a distance of 1.2 to 1.5 times Earth's radius is about a year. The lifetime of the same electrons is reduced to days and months at a distance of 1.5 to 2.5 times Earth's radius. At even greater distances, the lifetime of the particles is measured in minutes. Because Earth is strongly influenced by the Sun's magnetic field, Earth's geomagnetic field does not decrease indefinitely with increasing distance. The solar wind pushes Earth's magnetic field and is deflected by it. At about 10 Earth radii, the radiation belt ends abruptly.

Particles of trapped radiation may be lost in two ways. During a magnetic storm, the magnetosphere may lose or gain particles. This occurs at distances of 1.0 to 1.5 times Earth's radius. The other mechanism occurs at distances greater than 8 times Earth's radius. Small, rapid variation in the magnetic field at such distances scatters trapped particles, dumping them into the atmosphere. In a similar fashion, it is seen that charged particles in Uranus's magnetosphere are swept down into the planet's upper atmosphere by collisions with particles in its ring system.

Beautiful auroral displays occur when the charged particles are dumped into Earth's upper atmosphere. Solar flares eject into space streams of high-energy protons and electrons. When these beams of high-energy particles are directed toward Earth, Earth's magnetic field is partially disrupted. Particles trapped within the field lines can escape downward toward Earth at the lower ends of the radiation belts. High-energy particles, reinforced with

particles from the Sun, energize the upper atmosphere, causing luminous and often colorful auroras.

Knowledge Gained

The years 1957 and 1958 were designated the International Geophysical Year (IGY), an international scientific tour de force to advance understanding of Earth sciences. As contributions to IGY, the Soviet Union and the United States both pledged to place a satellite in orbit about the Earth. Russia's Sputnik 1 beat the American effort to orbit. However, the American effort was the first to gather useful scientific information. With the data returned by Explorer 1, America's first artificial satellite, a high-energy radiation belt was detected by James A. Van Allen and his assistants, George H. Ludwig, Carl E. McIlwain, and Ernest C. Ray. The same observations were made by Explorer 3, launched by the U.S. Army on March 26, 1958, and Sputnik 3, launched by the Soviet Union on May 15, 1958.

Later, a satellite was launched as part of Project Argus, which studied the location, height, and yield of electron blasts. This project was carried out by the Advanced Research Projects Agency. The belt of electrons produced by the Argus nuclear explosions developed at a distance of twice Earth's radius. Explorer 4, launched on July 26, 1958, carried four Geiger counters to handle high levels of radiation. One of these Geiger counters was shielded with a thin layer of lead to keep out most of the radiation. The satellite reached a height of 2,200 kilometers and registered an intensity of high-energy radiation. From the data returned, scientists concluded that Earth is surrounded by belts of high-energy radiation consisting of particles originating from the Sun and trapped in the lines of force of Earth's magnetic field. These were named the "Van Allen radiation belts."

Explorer 4 obtained a kidney-shaped intensity contour of Earth's inner belt. Data from early Pioneer spacecraft suggested a solar origin of soft particles populating the outer zone. Three Pioneer probes and Luna 1 discovered the crescent-shaped intensity contours of the outer belt. Sputnik 3 data helped identify the bulk of the outer belt particles as low-energy electrons (10 to 50 keV).

Several more human-made belts were produced in 1962. The Starfish project, an American venture, created a belt much wider than the Argus belt. Decay of some of the particles took several years in low altitudes. In the same year, the Soviets created at least three similar belts. More sophisticated versions of the instrumentation used in these early probes of Earth were incorporated into spacecraft sent to other planets in the solar system. Probes

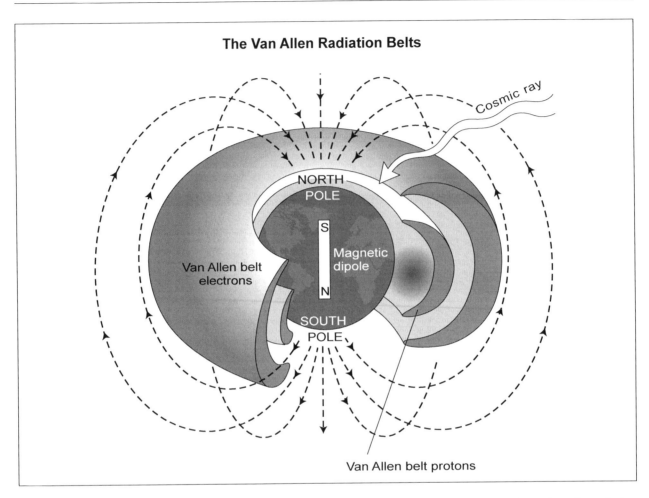

The Van Allen Radiation Belts

Cosmic ray

NORTH POLE

S

Magnetic dipole

N

Van Allen belt electrons

SOUTH POLE

Van Allen belt protons

to Mercury, Jupiter, Saturn, Uranus, and Neptune have discovered radiation belts similar to Earth's Van Allen belts. A planet needs a magnetic field to trap charged particles into a radiation belt or system of radiation belts. For that very reason, there are no significant belts of trapped charged particles at Venus or Mars.

Context

In the earliest days of space exploration, gauging the intensity of Earth's radiation belts with uncrewed spacecraft was crucial as a first step toward sending humans into space. Both the United States and the Soviet Union had a vested interest in the results of early investigations of the magnetosphere.

The Van Allen belts, while lifesaving in that they keep dangerous radiation from reaching the surface of the planet, are potentially hazardous to Earth-orbiting spacecraft. They threaten electronics systems and instrumentation and can interfere with radio transmissions. In the late

1950's, it was not known just how hazardous the radiation surrounding Earth would prove to humans. While it would not be wise to base a space station within the radiation belts, the belts themselves pose little threat to humans, who quickly punch through on voyages of exploration beyond the Earth; that was the case back in the Apollo program and will be true of the National Aeronautics and Space Administration's (NASA's) planned Constellation program flights to the Moon and later to Mars. However, leaving the safety of orbit beneath the Van Allen belts does expose astronauts to the potential hazards of ionizing radiation streaming outward into the solar system from the Sun. Therefore, special protection must be provided to humans on expeditions beyond low-Earth orbit.

The relationship between auroras and the Van Allen belts has been studied for decades, but although the overall phenomenon has been well characterized, all is not completely understood. Scientists do know that most bright auroras are produced by electrons dumped into

Earth's atmosphere by solar flares. The auroral particles are the electrons escaping from the outer Van Allen radiation belt. The average kinetic energy of the electrons is 32 keV. The leakage of corpuscular radiation into the auroral zones is the most important loss of corpuscular radiation from the outer Van Allen belt.

Satya Pal

Further Reading

Bone, Neil. *The Aurora: Sun-Earth Interaction.* New York: John Wiley, 1996. One volume in the Ellis Horwood Library of Space Science and Space Technology. Devoted to describing the electrodynamics of the Sun-Earth environment that produce auroral displays.

Bothmer, Volker, and Ioannis A. Daglis. *Space Weather: Physics and Effects.* New York: Springer Praxis, 2006. A selection from the publisher's excellent Environmental Sciences series, this is an overview of the Sun-Earth relationship and provides a historical and technological survey of the subject. Projects the future of space weather research through 2015 and includes contemporary spacecraft information.

Foerstner, Abigail. *James Van Allen: The First Eight Billion Miles.* Iowa City: University of Iowa Press, 2007. An engaging portrait of the legendary physicist, discussing his contributions to the World War II effort as well as to the advancement of studies of Earth's geomagnetic environment, his early efforts to study cosmic rays using balloon-launched rockets, the Explorer 1 story, and Van Allen's continuing participation in studying space physics until his passing in 2006.

Gregory, Stephen A. *Introductory Astronomy and Astrophysics.* 4th ed. San Francisco: Brooks/Cole, 1997. Suitable as a textbook for introductory college courses or advanced high school courses in general astronomy. Covers all topics from solar-system bodies to cosmology. Some errors and issues with mathematical presentations.

Kallmann-Bijl, Hildegaard, ed. *Space Research: Proceedings of the First International Space Science Symposium.* New York: Interscience, 1960. A detailed explanation of the theory behind the nature, origin, and composition of the inner, outer, and third radiation belts is provided. Dated and technical, but provides a useful historical perspective.

Milone, Eugene F., and William Wilson. *Solar System Astrophysics: Background Science on the Inner Solar System.* New York: Springer, 2008. Rigorous and highly mathematical presentation involving geophysics, atmospheric physics, and mineralogy covering all aspects of planetary science. Includes results from the Mars Exploration Rovers and Cassini spacecraft.

Moldwin, Mark. *An Introduction to Space Weather.* Cambridge, England: Cambridge University Press, 2008. This text introduces space weather, the influence the Sun has on Earth's space environment, to the nonscience reader. Discusses both the scientific aspects of space weather and issues of technological and societal import.

Savage, Candace. *Aurora: The Mysterious Northern Lights.* New York: Firefly Books, 2001. Heavily illustrated with photographs of auroral displays, this book provides a history of scientific investigation of auroral phenomena.

Sullivan, Walter. *Assault on the Unknown.* New York: McGraw-Hill, 1961. Sullivan describes the discovery of the Van Allen belts in great detail.

Van Allen, James A. *The Magnetospheres of Eight Planets and the Moon.* Oslo: Norwegian Academy of Science and Letters, 1990. A technical summary of all major magnetic structures in the solar system, written by the prolific researcher after whom the Van Allen belts are named.

_____. "Radiation Belts Around the Earth." *Scientific American* 200 (March, 1959): 39-47. A seminal article by the discoverer of the Van Allen belts. Well illustrated. Provides insight into the nature of the original discovery and the early days of the space race.

VENUS'S ATMOSPHERE

Categories: Planets and Planetology; Venus

The atmosphere of Venus has a surface temperature of about 743 kelvins and a surface pressure of about 90 Earth atmospheres. Its clouds consist largely of carbon dioxide, and droplets of sulfuric acid rain down to the surface.

Overview

The second planet from the Sun and Earth's immediate inner neighbor, Venus is often called Earth's twin, because the masses and radii of the two planets are very similar. Venus's mass is 82 percent that of Earth. Its radius is only 5 percent less than Earth's. Under ordinary circumstances, no objects in the sky other than the Sun and the Moon surpass Venus in brightness. Viewing Venus with one or more of the other planets also visible in the sky shortly before sunrise or briefly after sunset can be an awe-inspiring sight. In ancient times, since Venus could at times be seen in the morning skies and at other times in the evening skies, the planet was actually thought to be two different objects; they were given the names Phosphoros and Hesperus for the morning and evening star, respectively.

Venus revolves once around the Sun every Earth 224.7 days. It rotates on its axis once every 243.01 days. The inclination of its orbit is about 3.5° with respect to the ecliptic plane. Its orbit, like those of all the other planets, is an ellipse, but it is very close to being a perfect circle. The tilt of Venus's axis of rotation is about 17.8°, as compared with the Earth's 23.5°. As a result, any seasonal changes of weather on Venus would be less extreme than on Earth. Remarkably, its rotation is retrograde, backward, as compared with the direction of revolution, or clockwise, as seen from above the ecliptic plane. It is not known with certainty why Venus has a retrograde rotation. Virtually all other objects in the solar system rotate and revolve prograde, or counterclockwise as seen from above the ecliptic.

At each inferior conjunction (each time Venus and the Sun are aligned in the sky with Venus closer to Earth than to the Sun), the same face of Venus points toward the Earth. This phenomenon may mean that Venus's rotation is influenced by the Earth's gravitational pull; however, some information indicates that the alignment at inferior conjunction is not exactly perfect and therefore may be coincidental.

As seen from Earth, Venus's maximum disk size, or angular diameter on the sky, is about 0.02°, and its minimum angular diameter is about 0.003°. The maximum or minimum size on the sky corresponds to the closest and farthest distances from Earth. In contrast, the Sun and the Moon have an apparent angular diameter of about 0.5°. Given these observational circumstances, little can be learned about the planet's atmosphere from telescopic observations. The planet's average density is 5.25 grams per

cubic centimeter, and the surface acceleration caused by gravity is 0.903 times that of Earth. Escape velocity from Venus's surface is 10.3 kilometers per second. That means Venus can retain its atmosphere virtually indefinitely, since most of the molecules at the top of the atmosphere will travel at speeds well under 10.3 kilometers per second.

Through a telescope, an image of Venus is somewhat disappointing; only the planet's phases are obvious. Exceptional atmospheric and viewing conditions at good astronomical sites have allowed scientists to see and photograph subtle variations in shading on bright cloud tops of the planet's lit side. However, no part of the surface of Venus is visible through telescopes, because of its thick cloud cover.

Sulfur is probably released into the atmosphere of Venus by outgassing volcanic processes. Sulfur rises and bonds with water and oxygen to produce sulfuric acid. It appears as a fairly thick haze and is more strongly concentrated than the acid found in the battery of an automobile. These clouds are less murky than fog, with visibilities

This image of one complete Venusian hemisphere is a composite of high-resolution Magellan images whose gaps are filled by images from the Arecibo radio telescope in Peru. (NASA/JPL/USGS)

within them of perhaps several hundred meters. Top layers of these clouds are about 80 kilometers above the planet's surface. The tops of the sulfuric acid-laden clouds have winds that move faster than 300 kilometers per hour, a speed comparable to that of the Earth's jet streams. These high winds swirl around the planet in about four days. Circulating motions cause gas to rise near the equator and descend near the poles, probably a direct consequence of excess solar heating in the equatorial region.

The atmosphere of Venus is mostly carbon dioxide. Venus has undergone a spectacular greenhouse effect, caused by particles in the clouds trapping or absorbing infrared radiation. Most sunlight is reflected by clouds back into space. The albedo of Venus is about 0.8; that is, 80 percent of incident radiation striking the cloud tops is reflected back into space. The 20 percent that penetrates the clouds warms the surface sufficiently to heat the surface rocks and terrain. These surface structures radiate, essentially at infrared wavelengths, and that radiation cannot penetrate the clouds and gets trapped. Heat builds, and the high temperature releases gases (mainly carbon dioxide) from rocky minerals into the atmosphere, which in turn drives the greenhouse effect further. A cycle develops: The carbon dioxide traps the radiation, which heats the surface and atmosphere, which then triggers the release of more carbon dioxide. This continues until the entire atmosphere reaches a sufficiently high temperature (740 kelvins) to radiate as much energy back into space as it receives.

High above the clouds, the atmospheric layers called the exosphere consist mostly of hydrogen and helium. These strata are affected by intense incoming ultraviolet solar radiation. The radiation ionizes atoms, making them electrically charged. The uppermost layers form the ionosphere. Venus's ionosphere is not as intensely ionized as Earth's.

Venus's surface is quite flat compared with that of the Earth; however, there are at least three major elevated regions, and they exhibit features that influence the atmosphere. The largest, Ishtar Terra, is similar in size to Australia. The others are about the size of the largest islands in Indonesia. Many less elevated regions are also present. On Ishtar Terra, a mountain called Maxwell Montes is apparently higher than Mount Everest with respect to the surrounding flat terrain. Large volcanoes are present on each of the three plateaus. Volcanoes imply gas release from the planetary interior; therefore, Venus's atmosphere is attributable in part to volcanic outgassing. Although the atmosphere is composed mostly of carbon

dioxide, there is some nitrogen and sulfur dioxide, very little water vapor, and trace amounts of various other gases. Sulfur dioxide is outgassed by volcanoes on the Earth but is quickly diluted by rain and moisture. In contrast, sulfur dioxide outgassed in the dry atmosphere of Venus is very stable, accounting for the efficient production of sulfuric acid in the clouds and elsewhere.

Continent-building processes caused by plate tectonics on the Earth may have occurred on Venus, though to a far lesser extent. This idea is based mostly on the fact that only a few elevated regions exist. Therefore, the outgassing brought on by a variety of volcanic actions related to plate tectonics is probably slower and less effective on Venus, compared with the heat-induced gases, such as carbon dioxide, from rocks.

One other feature of Earth's atmosphere that probably does not exist to any extent on Venus is production of high-level auroras (or, as they are known on Earth, the northern and southern lights), because Venus has little or no magnetic field. Since its rotation rate is very slow, one would expect a weak but nevertheless measurable magnetic field. Several theories have been put forth to explain this lack. One theory is that, like the Earth, Venus undergoes polarity changes of its overall dipolar magnetic field. The Earth's magnetic field is explained by the dynamo hypothesis. The rotating core produces loop currents in the heated molten regions, which in turn produce a magnetic field. Geological and paleontological evidence suggests that the Earth's magnetic field undergoes reversals of polarity at irregular intervals. During the reversal periods, little or no field is present. It is possible that Venus could be undergoing such a magnetic reversal phase. In fact, either it is in such a phase or the dynamo hypothesis is incorrect.

Methods of Study

The first attempt to send an interplanetary probe to Venus did not fare well. Mariner 1 launched on July 22, 1962. A software error caused its booster to veer dangerously off course while low in the atmosphere, and it was destroyed on purpose by the range safety personnel at Cape Canaveral.

Mariner 2, a sister spacecraft to the failed first American Venus probe, launched successfully on August 27, 1962. Fortunately, this probe was able to fly within 34,833 kilometers of the Venusian surface on December 14, 1962. Although it carried no photographic equipment, Mariner 2 provided a treasure trove of new information about the shrouded planet. The spacecraft was outfitted with Geiger tubes, an ion chamber, a cosmic dust detector, a microwave radiometer, and a magnetometer experiment. Mariner 2 determined that the Venusian surface temperature was more than 670 kelvins. It detected neither a planetary magnetic field nor any Van Allen-like radiation belts about the planet. It continued to collect data about particles and fields in interplanetary space until contact was lost on January 3, 1963, at a distance of 87 million kilometers from Earth.

Mariner 5 was launched on June 14, 1967, and was sent to fly by Venus. This spacecraft also did not include photographic or television cameras. It was equipped with radio science and ultraviolet experiments as well as particle and magnetic field detectors. Mariner 5 encountered Venus on October 19, 1967, coming within 4,000 kilometers. The spacecraft investigated Venus's cloud tops and the solar wind interacting with interplanetary magnetic fields.

Early images taken by the space probes Mariner 10 and Pioneer Venus in reflected solar ultraviolet light revealed considerable variation in shading on Venus. Variation is caused by radiation coming from different levels of the clouds. The study of the motion of these clouds has indicated that the top layers can rotate at very high speeds, approaching perhaps 100 kilometers per hour. These high-strata, rapid wind velocities are in part caused by the hot, sunlit clouds transferring heat to the colder dark side. The cloud structure shows three distinct strata: a high, thick layer; a medium-high haze layer; and a lower, medium-thick layer. From this lower layer downward it is essentially clear all the way to the surface.

The Pioneer Venus atmospheric probes, sent to the surface via a combination of small retrorockets and parachutes, and Soviet Venera landers measured a decrease in the wind velocity at lower altitudes. On the surface, winds are essentially gentle breezes. Heat transfer and exchange in the atmosphere are very dependent on the density and pressure of various layers. Lower levels of the atmosphere are under superhigh pressures. Soviet Venera landers and the Pioneer space probes measured pressures of about 90 atmospheres at the surface. Heat transfer is so efficient that there is no large-scale difference between daytime and nighttime temperatures at the surface. Mariner 10 and Pioneer Venus ultraviolet images indicated an overall circulation pattern: Atmospheric gas rises at the equator and descends at the poles. With slow rotation of the planet, this circulation pattern is highly stable.

Four Venera landers managed to set down on the surface, perform experiments, and obtain electronic images of the surroundings using a fish-eye lens or wormlike view of the terrain to the horizon. At both landing sites, rocks and the horizon in the clear atmosphere are visible. Rocks are clearly of volcanic origin. In some cases their sharp edges indicate little or no erosion, suggesting recent volcanic origin. One would expect that erosion, under the high pressure and intense heat, would quickly deform and erode the rocks' edges. Pioneer Venus detected the existence of sulfuric acid droplets, which at the high temperatures and pressures present is very corrosive.

Pioneer Venus was equipped with a radar ranger. Scientists could send radar beams to the surface of Venus from the spacecraft and measure the time interval from the emission of the radar to the subsequent receiving of the reflected echo. This procedure allowed accurate measurement of distances between the spacecraft and the surface. After a compilation of such observations, the Pioneer Venus mission team was able to provide a detailed map of the surface terrain for the first time. Better maps would have to await a more sophisticated radar system placed in orbit about Venus.

Vega 1 and 2 were ambitious missions involving identical carrier spacecraft that each delivered both a lander based on the Venera design and an instrumented balloon to Venus before both carrier spacecraft were then redirected to join an international group of spacecraft intercepting and studying Halley's comet near its 1985/1986 perihelion passage. The carrier craft were outfitted with an imaging system, an infrared spectrometer, and a spectrometer capable of ultraviolet through infrared observations, detectors of dust and micrometeoroids, a plasma energy analyzer, a magnetometer, wave and plasma analyzers, a neutral gas mass spectrometer, and an energetic particle analyzer. The Vega 1 and 2 carrier craft encountered Venus on June 11 and 15, 1985, respectively, having several days earlier ejected their lander and balloon payloads. Neither carrier craft provided deep new insights into the nature of the Venusian atmosphere, but they did set the stage for the unique balloon payload and their results.

Venus's atmosphere apparently provided a particularly strong wind gust when the Vega 1 lander was still 20 kilometers above the planet's surface. This even activated the surface experiments early, and no results were produced after touchdown. Vega 2 operated properly and on June 15, 1985, safely touched down in the eastern Aphrodite Terra region. This lander determined the local atmospheric pressure to be 91 atmospheres, with a surface temperature of 736 kelvins. It endured the extreme environment for just under an hour, but, before it failed, the lander determined a rock sample to be a variety of anorthesite.

The Vega 1 and 2 balloons floated in the planet's atmosphere, providing data for about forty-six hours at an altitude of approximately 54 kilometers. The balloons were small in mass and size (25 kilograms, about 55 pounds, and 3.4 meters, or 11 feet, in diameter) but were able to dangle a gondola assembly filled with instruments to sample and measure the Venusian atmosphere. The balloons began their mission after being deposited on Venus's dark side. They sank to a depth of 50 kilometers before rising again to an altitude of 54 kilometers where they determined the pressure and temperature to be similar to those conditions on Earth. However, wind speed was nearly that of hurricane status, and at this altitude the carbon dioxide-rich atmosphere had a strong concentration of sulfuric acid with far less hydrofluoric and hydrochloric acid. Before losing electrical power, the balloons registered a variable vertical component to the atmospheric winds upon which they floated. Also they survived long enough to move from the dark side to the planet's illuminated side.

The National Aeronautics and Space Administration's (NASA's) next probe to Venus was named after the great Portuguese explorer Ferdinand Magellan. Its goal was no less daunting than to use an imaging radar to map at least 98 percent of Venus's surface to a resolution of 100 meters or less. Magellan was deployed from the space shuttle *Atlantis* on the STS-30 mission on May 5, 1989, and sent on its way toward the inner solar system. The spacecraft arrived in Venus orbit on August 10, 1990. Its synthetic aperture radar system was able to peer through the thick atmosphere. After four years of mapping, radar altimetry, and gravitational field measurements, NASA intentionally drove Magellan through the planet's atmosphere on October 12, 1994. In one final experiment, information was inferred about the atmosphere as the spacecraft heated up, and contact was eventually lost when Magellan was destroyed.

The European Space Agency (ESA) launched the Venus Express spacecraft on November 9, 2005. Largely a twin of ESA's successful Mars Express spacecraft, but modified to study Venus, Venus Express entered a nine-day-period polar orbit about Venus on April 11, 2006. Then for science operations to commence, that orbit was altered to have a twenty-four-hour period. Equipped with a penetrating radar, Venus Express began generating a surface map of the planet at resolutions even better than Magellan had achieved. However, science objectives of Venus Express also included detailed atmospheric studies. Aboard the spacecraft were infrared, visible-spectrum, and ultraviolet instruments to observe Venusian atmospheric characteristics and determine temperature profiles as a function of altitude.

Venus's Atmosphere Compared with Earth's

	Venus	*Earth*
Surface pressure (bars)	92	1.014
Surface density (kg/m³)	~65	1.217
Avg. temperature (kelvin)	737	288
Scale height (kilometers)	15.9	8.5
Wind speeds (meters/second)	0.3-1.0	up to 100
Composition		
Argon	70 ppm	9,430 ppm
Carbon dioxide	96.5%	350 ppm
Carbon monoxide	17 ppm	—
Helium	12 ppm	5.24 ppm
Hydrogen	—	0.55 ppm
Hydrogen chloride	tr	—
Hydrogen fluoride	tr	—
Krypton	—	1.14 ppm
Neon	7 ppm	18.18 ppm
Nitrogen	3.5%	78.084%
Oxygen	—	20.946
Sulfur dioxide	150 ppm	—
Water	20 ppm	1%
Xenon	—	0.08 ppm

Note: Composition: % = percent; ppm = parts per million; tr = trace amounts.
Source: Data are from the National Space Science Data Center, NASA/Goddard Space Flight Center.

Venus Express discovered a rather unexpected double vortex feature located around the south pole of the shrouded planet. This remarkable find occurred on the spacecraft's very first highly elongated orbit about Venus. Thus Venus Express was able to examine the planet's atmospheric patterns in ultraviolet and infrared from a global perspective (when far from Venus) and at close range (as it approached its low point in orbit). A vortex feature had been seen previously over the planet's north pole by earlier spacecraft, but a double vortex with a stable structure was quite unusual. Invoking the high wind speed of the upper atmosphere and the convection of rising hot air was insufficient to explain this double vortex. Using infrared sensors, Venus Express was able to map out windows in the atmosphere through which thermal radiation could escape to space. That modeling assisted scientists in determining cloud structures as a function of altitude above the tremendously hot planetary surface.

Some earlier probes had provided circumstantial evidence that lightning was present in Venus's atmosphere, but others produced data strongly suggesting there was a total lack of lightning. In 2006, the magnetometer aboard Venus Express provided definitive data that lightning does occur in Venus's atmosphere.

Venus Express made another important discovery in 2008, detecting hydroxyl molecules. This was the first time on a planet other than Earth that hydroxyl molecules had been detected; hydroxyl is a molecular ion consisting of one oxygen and one hydrogen atom bonded together covalently. Hydroxyl was detected at an altitude of 100 kilometers above the Venusian surface, which Venus Express accomplished by means of the Visible and Infrared Thermal Imaging Spectrometer, picking up the faint infrared light emitted by these molecules in a very narrow band of Venus's atmosphere. That band appears to be only 10 kilometers thick.

Hydroxyl has been found around comets. However, planetary atmospheres produce the molecule in a very different manner from that involved in comets. Hydroxyl on Earth is associated with the abundance of ozone in the upper atmosphere. Thus, detection of hydroxyl in Venus's atmosphere suggests that Venus still retains some Earth-like aspects. Absorption of ultraviolet light by hydroxyl molecules is important to the heating balance of any planetary atmosphere. The hydroxyl data from Venus Express would greatly assist planetary scientists in fine-tuning their models of the Venusian atmosphere.

Finally, in late 2008 Venus Express for the first time detected water being lost from Venus's daylight side. The previous year, this spacecraft's Analyzer of Space Plasma and Energetic Atoms detected the signature of hydrogen being stripped away from the planet's nightside. The orbiter's magnetometer was used to find hydrogen dissociated from water coming off the daylight side to be lost into space. Solar wind particles penetrate Venus's atmosphere, since the planet lacks a protective magnetic field. Scientists believe that solar wind particles break water molecules into two parts hydrogen and one part oxygen. Oxygen and hydrogen have been found escaping the nightside in the right proportion, but oxygen escaping from the daylight side was not seen in the 1:2 ratio required if the hydrogen seen comes from water. In any event, the solar wind mechanism was believed by many to be the means whereby, over time, Venus lost much of its original water.

Context

Earth's atmosphere has the potential to become more like that of Venus. There are two ways that such a situation could develop. If Earth moved closer to the Sun, increased solar heat would release more carbon dioxide into the atmosphere. Limestone rock (calcium carbonate) and dissolved carbon dioxide in the oceans provide a tremendous store of trapped carbon dioxide. Under higher temperatures, the rocks and seashells would chemically release carbon dioxide, and the greenhouse mechanism would raise the atmosphere's temperature. As a result, new carbon dioxide would be released and would speed the greenhouse mechanism. The second way that Earth's atmosphere could become more like that of Venus involves pollution of the atmosphere to the extent that enough carbon dioxide accelerates the existing greenhouse effect.

More generally, study of Venus's atmosphere helps scientists to better understand terrestrial weather and climate. Earth has an atmosphere composed mostly of nitrogen. The weather, however, is influenced primarily by water molecules and carbon dioxide molecules. These substances are found in Earth's atmosphere only in trace amounts; nevertheless, they are responsible for most of the heat transfer around the globe. On Venus, the weather is controlled by the atmosphere's main constituent, carbon dioxide. The contrasts between weather processes on Venus and those on the Earth have led to a more complete understanding of the latter. In an even larger context, comparative planetology studies of Venus, Earth, and Mars contribute to a better understanding of Earth's complex weather system and atmospheric physics. Moreover, such study helps us learn why three planets, all within the Sun's habitable zone, could evolve so differently. To understand Earth's evolution fully, it is necessary to know why the Venusian atmosphere became thick in carbon dioxide at great pressure (so that a planetary greenhouse effect led to runaway temperatures), and also to understand why the Martian atmosphere became thin in carbon dioxide at low pressure and low temperature. Earth, on the other hand, developed a nitrogen-oxygen atmosphere with traces of carbon dioxide; this led to reasonable temperatures and pressures and the development of a complex biosphere and interactive oceanic-atmospheric processes to maintain a dynamic equilibrium.

James C. LoPresto and David G. Fisher

Further Reading

Beatty, J. Kelly, Carolyn Collins Petersen, and Andrew Chaikin, eds. *The New Solar System*. 4th ed. Cambridge, Mass.: Sky, 1999. Amply illustrated with color images, diagrams, and informative tables, this book is aimed at a popular audience, but it can also be useful to specialists. Contains an appendix with planetary data tables, a bibliography for each chapter, planetary maps, and an index.

Cattermole, Peter John. *Venus: The Geological Story*. Baltimore: Johns Hopkins University Press, 1996. Provides a comprehensive presentation of the latest understanding of Venus based on Magellan data.

Elkins-Tanton, Linda T. *The Sun, Mercury, and Venus*. New York: Chelsea House, 2006. Examines the innermost portion of the solar system and the star, our Sun, which plays such a prominent role in the evolution of both planets. For the general audience with an interest in science.

Esposito, Larry W., Ellen R. Stofan, and Thomas E. Cravens, eds. *Exploring Venus as a Terrestrial Planet*. New York: American Geophysical Union, 2007. A collection of articles covering all major areas of planetary research on Venus. Technical.

Fimmel, Richard O., Lawrence Colin, and Eric Burgess. *Pioneering Venus: A Planet Unveiled*. Washington, D.C.: National Aeronautics and Space Administration, 1995. A complete summary of the findings of Pioneer Venus as of 1983. Pioneer Venus orbited Venus and sent several probes into the atmosphere of the planet. It also mapped the planet using a radar-ranging device.

Freedman, Roger A., and William J. Kaufmann III. *Universe*. 8th ed. New York: W. H. Freeman, 2008. A college text on astronomy, somewhat more advanced than many introductory texts, but with a wealth of detail and excellent diagrams. Comes with a CD-ROM. The chapter on Venus is lucid and filled with spectacular diagrams and photographs.

Grinspoon, David Harry. *Venus Revealed: A New Look Below the Clouds of Our Mysterious Twin Planet*. New York: Basic Books, 1998. A thorough examination of the geology of Venus, incorporating Magellan mapping and other data. Explains the Venusian greenhouse effect. A must for the planetary science enthusiast who wishes to read an integrated approach to science and history. Includes speculation about Venus's past.

Hartmann, William K. *Moons and Planets*. 5th ed. Belmont, Calif.: Thomson Brooks/Cole, 2005. An updated version of a classic text that covers all aspects of planetary science. A comparative planetology approach is used rather than presenting just one chapter on all characteristics of Venus.

Marov, Mikhail Ya, and David Grinspoon. *The Planet Venus*. New Haven, Conn.: Yale University Press, 1998. Marov was Soviet Venera mission chief scientist, Grinspoon a NASA-funded scientist studying Venus. Together they provide a coordinated description of American and Soviet attempts to learn the secrets of Venus, a planet shrouded in mystery. For both general readers and specialists.

Morrison, David, and Tobias Owen. *The Planetary System*. 3d ed. San Francisco: Pearson/Addison-Wesley, 2003. A textbook at the beginning college level, introducing the scientific knowledge of the solar system as of 1988. The chapter on Venus goes into detail about the greenhouse effect and the contrasting atmospheres of Venus and the Earth.

Spanenburg, Ray, and Kit Moser. *A Look at Venus*. New York: Franklin Watts, 2002. A look beneath the thick clouds of Venus written for a younger audience.

VENUS'S CRATERS

Categories: Planets and Planetology; Venus

Impact craters are the most numerous and most easily recognized surface features in the solar system. Because of their pristine nature, Venusian impact craters provide a unique opportunity for astrogeologists to study the effects of atmospheric variabilities and gravity in the formation of planetary surfaces.

Overview

One of the earliest and most crucial phases in the early development of the solar system was the Great Bombardment. During this phase, planetary surfaces were under intense bombardment by cosmic debris. For the Earth and Moon, the Great Bombardment began about 4.6 billion years ago and declined about 3. 8 billion years ago. Because the Moon has no active tectonism or atmospheric processes, impact craters there are mostly scars left from this time of intense bombardment. Because Earth, by contrast, is tectonically active and continuously subjected to weathering processes, craters of such ancient ages are not visible on Earth's surface and have been identified only on the cratons of continents, which undergo relatively little resurfacing from weathering, tectonism, volcanism, or other processes. The majority of impact craters identified on Earth, therefore, are geologically young and do not represent impacts from the Great Bombardment. Venus, Earth's closest planetary neighbor, presumably was also subjected to this intense cosmic bombardment, but it appears that Venus totally lacks the ancient, heavily cratered surface that occurs on Mercury, Mars, the Moon, and the rocky moons of other planets.

The detailed morphology of the surface of Venus is known through research done on remote-sensing data obtained from successful missions mounted by the National Aeronautics and Space Administration (NASA)—the Mariner (1962-1975), Pioneer (1978), Galileo (1990), Magellan (1990-1994), Cassini-Huygens (1998-1999), and MESSENGER (2004) uncrewed spacecraft missions—as well as the Soviet Union's Vega (1985) and Venera missions (1967-1984) and the European Space Agency (ESA) Venus Express (2005) mission. Other data have been accumulated through study of high-resolution images from Earth-based radar. The Magellan spacecraft, inserted into Venus orbit in 1990, yielded radar images with a resolution of a few hundred meters covering nearly 98 percent of the planet. In addition, the Soviet Venera missions mapped nearly 25 percent of Venus with additional radar imaging. These data have resulted in the mapping of nearly 100 percent of Venus and suggest that Venus lacks the ancient, heavily cratered surface occurring on other terrestrial planets and rocky satellites. The Venus Express mission has provided valuable data on the atmosphere of Venus and evidence suggesting that past oceans may have existed on the Venusian surface. A new Venus mission, Planet-C is planned for 2010 by the Japan Aerospace Exploration Agency (JAXA).

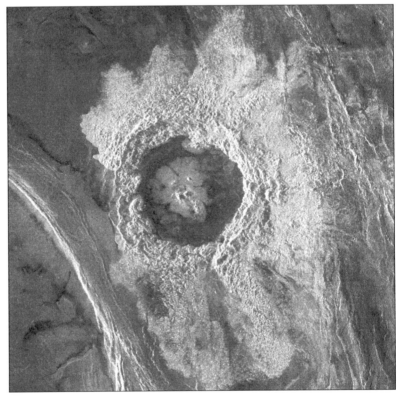

Venus's Dickinson Crater, 69 kilometers in diameter, from the Magellan spacecraft. (NASA/JPL)

Astrogeologists use the density of impact craters to determine the age of planetary surfaces. The older the surface, the more impact craters it will have accumulated over time. On Venus this dating technique is problematic, because there are relatively few impact craters. Based on the density of Venus's impact craters larger than 30 kilometers in diameter, estimates of cratering rates scaled for other terrestrial planets and rocky satellites, and the known population of asteroids crossing Venus's orbit, the planet's average surface age is estimated at between 450 million and 250 million years old, the younger age being more likely. This suggests the surface terrain of Venus may be less than 5 percent of the age of the solar system. However, these ages are average estimates, and based on superposition some Venusian impact craters, volcanic structures, and tectonic terrains are thought to be as young as 50 million years old. Average age estimates aside, for the last 700 million years Venus has been subjected to significant surface volcanism, which probably is the reason that so few impact craters are visible on the planet's surface: Older impact craters have been covered over with lava flows or destroyed during episodes of catastrophic volcanic eruptions.

Venus has the densest atmosphere of any terrestrial planet. The Venusian surface pressure is equivalent to 94 bars—more than ninety times the pressure humans feel from Earth's atmosphere (90 bars is approximately the weight of water at 1 kilometer below the surface of Earth's oceans). In addition, Venus's atmosphere is composed of 96 percent carbon dioxide and trace amounts of nitrogen, water vapor, argon, carbon monoxide, and other gases. These clouds seal in the Sun's heat, creating a perpetual greenhouse effect that boosts surface temperatures on Venus to around 753 kelvins. Clouds within the Venusian atmosphere are composed mainly of sulfuric acid and small amounts of hydrochloric and hydrofluoric acid. The presence of such a dense atmosphere effectively filters the numbers of potential impactors by severely decreasing their kinetic energy during transit and preventing all but the largest incoming objects from impacting the Venusian surface. Craters smaller than 1.5 kilometers appear not to exist on Venus. Many craters of this size are distinctly noncircular and form groups or clusters of craters. This phenomenon is attributed to the impactor's becoming fragmented as it passes through the dense Venusian atmosphere and hitting the surface like a shotgun blast rather than like an artillery shell.

The variety of morphologies seen in Venus's impact craters tends to depend on their size. As the diameter of Venusian craters increases, changes in crater morphology take place and appear to correlate directly with Venus's surface gravity and dense atmosphere. Much of the morphology of Venusian impact craters is unique. Craters larger than 11 kilometers in diameter exhibit morphological characteristics similar to comparable complex craters on other planets: a circular shape, surrounding ejecta blankets, well-defined rims, terraced walls, central ring structures or central peak complexes, and, in the largest craters, multiple-ring basins. However, smaller Venusian craters tend to display a wide variation in shape and structural complexity—the opposite of the cratering patterns seen on other terrestrial planets and rocky satellites.

The morphological divergence is most directly attributed to the greater atmospheric density of Venus. Large multiring basins on Venus display at least two, and sometimes three or more, rings and near-pristine morphology; are surrounded by blocky ejecta distributed in lobes or raylike patterns; and in some cases produce lavalike flows of ejecta traveling several radii from the crater. The ejecta patterns are attributed to the dense Venusian atmosphere's slowing the travel path and speed of debris exiting the crater during impact. Many of the largest Venusian impact craters appear to have little to no topographic relief. Shallowness of these craters may be linked to Venus's lower gravity, producing slower impact speeds, and the planet's high surface and crustal temperatures, producing a large volume of impact-generated melt that remains in a near-molten state, allowing it to flow over long periods of time and eventually fill in the crater.

It is also suggested large Venusian impacts could trigger the subsequent volcanic or tectonic activity that disguises, or eventually erases, them within the landscape. One of the more difficult aspects of studying Venusian craters is distinguishing impact craters from circular volcanic calderas. High-resolution radar images help in defining the morphology of impact craters versus volcanic features by distinguishing ejecta deposits from lava flows. Unfortunately, lavas generated from impact-triggered volcanism can complicate discerning these structures because the lava flows may infill the impact craters, making them look like calderas.

One of the most unusual phenomena associated with Venusian impact craters is parabolic halos. These halos surround about 10 percent of the youngest craters and usually expand westward. The halos are attributed to the formation of a pre-impact bow-shock wave created by the impactor's producing strong turbulence as it travels through the dense Venusian atmosphere. The turbulence lifts surface dust high into the air, and then

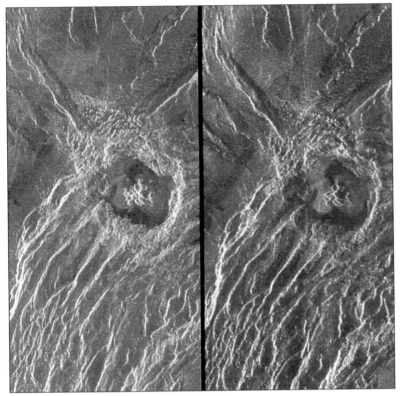

*A stereo image of Venus's Geopert-Meyer Crater from Magellan.
(NASA/JPL)*

prevailing easterly winds resettle the dust after the impact. Because the halos appear unaffected by volcanic, tectonic, or atmospheric processes, the haloed craters may be no more than 50 million years old, making them useful dating horizons.

Knowledge Gained

Identifiable impact craters on Venus are rare—slightly less than one thousand, or approximately 1 crater per million square kilometers—and large craters and basins are uncommon. Impact craters on Venus are randomly distributed and range in size from 1.5 to 270 kilometers in diameter. Venusian impact craters are unusual in that, almost without exception, they appear to be fresh, characterized by sharp rims and well-preserved ejecta deposits. This morphology suggests that the craters have not been subjected to significant erosional, volcanic, or tectonic activity. Only about 40 percent of Venusian craters appear slightly modified, 5 percent appear embayed by volcanic deposits, and 35 percent appear modified by tectonic activity. Venus's dense atmosphere filters out small meteors, so there is a lack

of small impact craters to chip away at larger craters. This situation favors the preservation of existing large craters. Furthermore, while there is currently no hydrogeologic cycle on Venus, there is evidence to suggest that there may once have been liquid oceans of some kind on the surface.

The pristine appearance of craters on Venus makes it appear the surface is both geologically young and of a relatively uniform age. This observation has significant implications for the geologic history of Venus. While resurfacing processes have most likely removed Venusian craters older that 450 million years, the morphology of existing craters is not what is expected for a steady balance between crater formation and crater loss caused by tectonic, volcanic, or erosional process. The unique observation is that Venusian craters of all ages look "fresh," suggesting that most of the present surface characteristics of Venus date from the end of a global resurfacing event that ceased about 450 million years ago.

Because of their lack of weathering, Venusian craters provide a unique opportunity for scientists to study the effects atmospheric variabilities and gravity have in forming planetary surfaces. While the total number of impact craters on Venus are not comparable to those on Mars, Mercury, and the rocky satellites, they do fall into morphological and age classifications similar to those of impact craters on Earth. Crater density and morphology suggest that cratering records of Venus and Earth are similar. Because the cratering data from these two planets are complementary, they provide interpretive guidelines for researching the roles that volcanism, tectonics, and erosional processes play in planetary resurfacing.

Context

The high temperatue and dense atmosphere of Venus slow incoming projectiles, destroying the smaller, high-velocity objects. This shielding effect influences the size of Venusian craters. Smaller craters appear to be absent on Venus because only large impactors can penetrate the Venusian atmosphere to reach the surface. It is estimated that as many as 98 percent of the craters between 1.5 and 35 kilometers in diameter that could have formed on Venus did not as a result of its dense atmosphere. Venus's high temperature and dense atmosphere also impeded the emplacement of ejecta during cratering by limiting flight distance and decelerating fragments, resulting in lobate ejecta blankets that are sharply defined and make up coarse blocks. Because of Venus's high surface temperature, rocks tend to be softer, less solid, and somewhat viscous. During an impact, these viscous rocks produce large amounts of impact melt, which works to fill the craters and make them topographically low. It is also suggested that large Venusian impacts could trigger regional tectonic or volcanic events by transferring their tremendous heat and shock energies into the planet's thin crust.

Randall L. Milstein

Further Reading

Esposito, L. W., E. R. Stofan, and T. E. Cravens, eds. *Exploring Venus as a Terrestrial Planet: Geophysical Monograph 176*. Washington, D.C.: American Geophysical Union, 2007. Addresses the open questions regarding Venus's geology, atmosphere, surface evolution, and future exploration. Includes results from the Venus Express mission.

Lopes, R. M., and T. K. P. Gregg. *Volcanic Worlds: Exploring the Solar System's Volcanoes*. New York: Springer, 2004. A general review of volcanic activity throughout the solar system. Comparisons are made between volcanic activity on Earth and other planets, showing how data from one planet can aid in the understanding of physical processes on another.

Spudis, P. D. *The Geology of Multi-ring Impact Basins: The Moon and Other Planets*. New York: Cambridge University Press, 1993. Although this is a technical book, the chapter on Venus is well illustrated and easy to read, with good comparisons to similar Earth structures.

Trefil, James S. *Other Worlds: Images of the Cosmos from Earth and Space*. Washington, D.C.: National Geographic Society, 1999. A richly illustrated book with exploration mission images of Venus and computer-generated three-dimensional perspectives of Venus's surface.

Uchupi, E., and K. Emery. *Morphology of the Rocky Members of the Solar System*. New York: Springer, 1993. Focuses on the morphology of planets and their satellites and the reasons for the differences and similarities between them. The book's theme is that the solar system should be approached as a single entity, not as a group of individual planets.

VENUS'S SURFACE EXPERIMENTS

Categories: Planets and Planetology; Venus

An understanding of the geology of the other planets in the solar system is important for understanding the geologic past and future of Earth. Venus holds many clues to this understanding, including its surface geology, which appears to be mostly igneous and basaltic in nature.

Overview

The planet Venus is considered one of the terrestrial or Earth-like planets because of its position in the solar system, its planetary diameter, its geology, and other characteristics. Despite being called Earth's twin because of those similarities, Venus is actually very different. The study of the planet Venus has been at best difficult because of its heavy cloud cover.

The best information on the Venusian surface and its soils came early from the Soviets, who focused on exploring the planet, successfully landing six spacecraft on the surface. Even though these craft operated for only limited amounts of time because of the planet's extreme temperatures and the pressure of its atmosphere, the data provided from them have given astronomers and geologists important clues to the soils on Venus. Much of the information obtained by the Soviets can be compared with that known about Earth, and to the data obtained from firsthand examination of the lunar rocks and soils. For example, photographs can be useful in examining the appearances of the soil, rocks, and their distribution. Images taken by Venusian landers can be compared with photographs of similar materials found on Earth and the Moon.

The first of the Soviet landers to provide clues to the Venusian soils was Venera 8, which made the first soft landing on Venus, on July 22, 1972, in a region generally thought to be like the rolling plains of Earth. The probe analyzed the surface and soils directly underneath it with a gamma-ray spectrometer designed to determine the chemical composition of surface material. Results showed that the soils under the Venera 8 were igneous, or volcanic, in origin. The layer was found to be approximately 4 percent potassium, approximately 200 parts per million uranium, and approximately 650 parts per million thorium. The layer was also determined to have a density of approximately 1.5 grams per cubic centimeter (in comparison, water has a density of 1 gram per cubic centimeter at a temperature of 277 kelvins). From these data, astronomers and geologists were able to ascertain not only that the soils under Venera 8 were igneous but also that they were probably similar to the granites or basalts found on Earth.

In 1975, Veneras 9 and 10 provided an even better look at the Venusian soils. Each lander transmitted an image that showed the soil and rocks surrounding it. These photographs revealed rocks that were on the average 20 centimeters wide, about 50 to 60 centimeters long, and slab-like in appearance. A few of the rocks showed evidence of volcanic origin. Many of the rocks had jagged edges, which demonstrates little erosion, although some did show signs of weathering. This relative lack of erosion surprised many astronomers and geologists. They had believed that, because of the planet's extremes in temperature, atmospheric pressure, wind velocity, and chemical composition, the photographs would show well-eroded landscapes. Astronomer Carl Sagan, among others, hypothesized that low wind velocities at the surface levels of Venus produce little effect on the rock. Apparently, the Venusian surface temperature stays fairly constant and thus does not create much wind. Chemical analysis of the rocks again showed the elements potassium, uranium, and thorium. Nevertheless, the sites differed in the type of rock material. At one Venera lander site, the rocks were basaltic in appearance, similar to those lining Earth's oceans. At the other site, the rocks were more like granite, similar to that found in Earth's mountains. The rocks appear to be relatively young in age. This would indicate that the planet has been geologically active in the geologically recent past. The Venusian soil in the areas observed photographically appeared to be loose, coarse-grained dirt. It was also evident from the photographs that Venus (or at least parts of it) is a dry and dusty

planet. Radar images from other Venera missions, as well as Pioneer Venus and the Magellan spacecraft, verified this for the rest of the planet.

The Soviets continued their studies of the Venusian surface with two additional spacecraft, Veneras 13 and 14. These two spacecraft performed similar examinations, but in a much more complex manner. Rather than single images, near-panoramic views of the landing sites were produced. Photographs showed rocks somewhat similar to those found at the Venera 9 and 10 landing sites. Rocks also showed evidence, however, of what appears to be thin layering, ripple marks, and fracturing, especially around Venera 14. Some rocks showed evidence of erosion. On Earth, rocks that show layering—such as sandstone and limestone—are usually sedimentary. Based on the photographs and measurements made by the spacecraft, several Soviet scientists suggested that the Venusian rocks might be sedimentary, but that has not been confirmed.

The possible cause or causes of the erosion remain unknown. In the absence of water, several possibilities have been suggested. These include chemical weathering or erosion caused by nearby volcanism and its resulting ash, dust, and lava. Chemical weathering seems the most likely explanation. Venus's thick atmosphere has cloud

Venus Compared with Earth

Parameter	Venus	Earth
Mass (10^{24} kg)	4.8685	5.9742
Volume (10^{10} km³)	92.843	108.321
Equatorial radius (km)	6,051.8	6,378.1
Ellipticity (oblateness)	0.000	0.00335
Mean density (kg/m³)	5,243	5,515
Surface gravity (m/s²)	8.87	9.80
Surface temperature (Celsius)	+450 to +480	−88 to +48
Satellites	0	1
Mean distance from Sun millions of km (miles)	108 (67)	150 (93)
Rotational period (hrs)*	−5,832.5	23.93
Orbital period	224.7 days	365.25 days

*The minus sign signifies a retrograde rotational period.
Source: National Space Science Data Center, NASA/Goddard Space Flight Center.

layers laced with sulfuric acid, which rains down as a caustic, corrosive agent on the surface.

Both spacecraft collected a cubic centimeter of Venusian soil for analysis. The probes utilized an X-ray source to stimulate emissions from the collected soil samples. This chemical analysis revealed that the samples were similar to basalt in composition, although the basalts differed at the two sites. Near the Venera 13 landing site, the type of basalt found is referred to as leucitic high-potassium basalt, while near the Venera 14 landing site, a tholeiitic basalt, similar to that found on the ocean floors on Earth, was found. The soil itself appeared fine-grained, and the photographs revealed many small rocks. It has been speculated that this also indicates that weathering processes of some type are at work, breaking down larger rocks into smaller ones, eventually reducing them to soil.

Another pair of Soviet probes, the Vega 1 and 2 spacecraft, landed on Venus in June, 1985. Vega 2 results revealed a Venusian soil and surface that are again similar to basalt. Nevertheless, the new data also revealed a surface rich in the element sulfur, which is usually associated with volcanism. This presence has provided another clue to the surface and geology of Venus.

Venera 8 landed about 5,000 kilometers east of an area referred to as the Phoebe region. Veneras 9, 10, 13, and 14 all landed between 900 and 3,000 kilometers east of the raised areas known as Beta Regio and Phoebe Regio. Even though the craft landed on and took samples from an area that could be of the same or similar geologic makeup, they have given astronomers and geologists a good idea of the planet's surface composition.

The probes produced mostly photographic data, although some chemical analysis was conducted on site. Thus, any discussion of soil samples is based on the evidence reported by these spacecraft, since no samples have ever been returned to Earth for detailed study. Spacecraft data have enabled astronomers and geologists to begin to understand not only the surface of the planet Venus and its chemical makeup but also the planet's evolutionary path.

Knowledge Gained

It appears that Venus may still be an active planet geologically, which scientists inferred from the discovery of high concentrations of sulfur in Venus's atmosphere. Thus, its soils, for the most part, must be considered with that fact in mind.

Analysis of Venusian rocks around the landers provided scientists with interesting but sometimes confusing data. For example, the fact that most of the rocks appear to lack signs of erosion at first seemed puzzling. An understanding of the weather patterns on Venus and the planet's atmospheric chemistry, however, has led to the development of theories relating the small-scale erosion to a low wind velocity at the surface because of its virtually uniform temperature. The rocks themselves appeared to be mostly igneous in nature. Most igneous samples appeared to be similar to basalt, much like those rocks and materials that line Earth's ocean floors. Some of the rocks resembled granite, like those that form Earth's mountains. However, despite apparent volcanic origins for Venus's crust, some specimens appeared to be sedimentary. This led to further questions which remain unanswered. Although the sedimentation process on Earth is usually accomplished by water, present-day Venus has no water, nor is there any evidence of water in its near past. The origins of this phenomenon remain unknown.

Analyzed samples varied slightly from site to site, as was expected by geologists, since samples on Earth also differ. In fact, variation of Earth samples is greater than that of the limited Venusian ones. Nevertheless, potassium—a key element in igneous and especially basaltic materials—was detected, as were uranium and thorium. Geologically, the rocks are relatively young, presenting additional evidence that Venus is a planet that may be experiencing continuous changes. Fine-grained soils were found at some sites, while coarser soils appeared at others. At one site, at least, smaller rocks led scientists to theorize that erosion does occur on the planet, thus producing soil.

Context

When the planets of the solar system are categorized, one usually finds two major groupings: the terrestrial or Earth-like planets; and the Jovian or Jupiter-like planets (sometimes also called the "gas giants"). Venus, because of its relative size, atmosphere, position within the solar system, and surface, is naturally among the set of terrestrial planets: Mercury, Venus, Earth, and Mars.

An understanding of the nature of the terrestrial planets, their atmospheres, planetary geologies, and soils can give astronomers and geologists clues to the pasts not only of these worlds but also of our own—revealing how these planets were formed, what geological changes they have undergone, and how they might be related. Venus holds many clues to the formation of the solar system. Unfortunately, observations of the Venusian surface are nearly impossible because of the dense atmosphere that surrounds the planet. Orbiting spacecraft provide information regarding the general geologic contours on the surface—the planet's mountains, valleys, craters, and other surface features—but hard evidence of the nature of the surface, particularly its soil, can come only from the surface of the planet. Prior to the landing of Soviet probes, no information about the Venusian surface existed.

Materials sampled and photographed in the vicinities of the Soviet landers proved to be mostly igneous in nature. Additional on-site chemical analysis showed these materials—both rocks and soils—to be similar to granite or basalt. Basalt-type materials are not unique to the second planet from the Sun. These materials have been found on Mars in the vicinities of the American Viking landers, in samples brought back from the Moon by the American Apollo crews, and by uncrewed Soviet spacecraft. As Soviet spacecraft became more sophisticated and knowledge of the harsh Venusian environment grew, landers were able to provide data on the surface of Venus, among other things. Additional information may provide scientists with clues to the past of the terrestrial planets, part of which is hidden in the Venusian surface and soil. Perhaps more important, Venusian soil information may provide clues to Earth's future, particularly regarding our planet's fragile environment.

Comparative planetology is essential for achieving a more complete understanding of our own planet. As Venus, Earth, and Mars started out relatively similar in the early solar system, and all three are in the habitable zone, why is it then that Venus is devoid of water and hot with a thick atmosphere of carbon dioxide, Earth is capable of supporting life, and Mars has no liquid surface water and is cold, with a thin atmosphere of carbon dioxide? Only when that question is answered will scientists have a clear idea of Earth's complex planetary environment.

Work performed along the way to that understanding has included spacecraft dispatched to the veiled planet Venus by both the Soviet Union and the United States. American spacecraft have only flown by or Venus or studied it from orbit. The National Aeronautics and Space Administration (NASA) has had more interest in exploring Mars than Venus. However, because the Soviets have had only bad luck when it comes to Martian exploration, they have emphasized the study of Venus with flyby craft, landers, orbiters, and even balloons temporarily floating within its hellish atmosphere. At the dawn of the twenty-first century, however, neither NASA nor the Russians had any plans to return to Venus for at least two decades. In the meantime, the European Space Agency's Venus Express began orbiting Venus on April 11, 2006,

conducting mapping operations and other scientific investigations of Venus's surface and atmosphere.

Mike D. Reynolds

Further Reading

Cattermole, Peter John. *Venus: The Geological Story.* Baltimore: Johns Hopkins University Press, 1996. A comprehensive presentation of the latest findings on Venus, based on Magellan data.

Corliss, William R., ed. *The Moon and the Planets.* Glen Arm, Md.: Sourcebook Project, 1985. A discussion of many solar-system phenomena that cannot be easily explained by prevailing scientific theories. Each anomaly is defined, substantiating data are presented, and the challenge the anomaly presents to astronomers is explained. Examples and references are also listed.

Esposito, Larry W., Ellen R. Stofan, and Thomas E. Cravens, eds. *Exploring Venus as a Terrestrial Planet.* New York: American Geophysical Union, 2007. A collection of articles covering all major areas of planetary research on Venus. Technical.

Fimmel, Richard O., Lawrence Colin, and Eric Burgess. *Pioneering Venus: A Planet Unveiled.* Washington, D.C.: National Aeronautics and Space Administration, 1995. A profusely illustrated scientific and technical publication from NASA that includes the Pioneer Venus data as well as a good deal of information from the Russian spacecraft dispatched to investigate Venus. Illustrations, bibliographic references, index.

Frazier, Kendrick. *Solar Systems.* Rev. ed. Alexandria, Va.: Time-Life Books, 1985. This text contains outstanding color photographs, diagrams, and coverage of the planets of the solar system.

Grinspoon, David Harry. *Venus Revealed: A New Look Below the Clouds of Our Mysterious Twin Planet.* New York: Basic Books, 1998. A thorough examination of the geology of Venus. Incorporates Magellan mapping and other data. Explains the Venusian greenhouse effect. Includes speculation about Venus's past. A must for the planetary science enthusiast who wants an integrated approach to science and history.

Hartmann, William K. *Moons and Planets.* 5th ed. Belmont, Calif.: Thomson Brooks/Cole, 2005. An updated version of a classic text that covers all areas of planetary science. The chapter on Venus covers all fundamental knowledge about the planet and spacecraft exploration of it.

Morrison, David, and Tobias Owen. *The Planetary System.* 3d ed. San Francisco: Pearson/Addison-

Wesley, 2003. The authors provide a detailed description of each of the planets and other bodies of our solar system. Some coverage of general astronomy and chemistry is included in introductory chapters. College level.

Snow, Theodore P. *The Dynamic Universe*. Rev. ed. St. Paul, Minn.: West, 1991. A general introductory text on astronomy. Covers historical astronomy, equipment used in astronomy, the solar system, stellar astronomy, galactic astronomy, cosmology, and life in the universe. The book features special inserts, guest editorials, and a list of additional readings at the end of each chapter. College level.

VENUS'S SURFACE FEATURES

Categories: Planets and Planetology; Venus

Enormous strides have been made in understanding the nature of the surfaces of the solid bodies in the inner solar system and the processes that have shaped them. Venus, however, has been a particularly difficult planet to study. The picture that is emerging suggests that Venus may be the only member of the four terrestrial planets, besides Earth, that remains geologically active.

Overview
Of all the terrestrial planets, Venus is the most similar to Earth in size and geologic composition. At 12,258 kilometers in diameter, it is only slightly smaller (by 511 kilometers) than Earth, and its density is within

2 percent of being identical. It lacks the polar flattening, equatorial bulge, and planetary magnetic field that Earth exhibits.

Geologic study of Venus is exceedingly difficult because of the fact that the planet is perpetually shrouded from view by thick clouds. Its surface has never been photographed from Earth-based telescopes or from spacecraft in orbit above the planet. However, in 1990 the Magellan spacecraft began generating high-resolution radar images of Venus's surface and thereby began producing detailed maps with a resolution of approximately 100 meters. A few panoramic photographs taken by several Soviet spacecraft have revealed the barren, rocky character of Venus's surface in the proximity of their landing sites. However, a very high surface temperature averaging 750 kelvins has limited the operating life spans of spacecraft that have landed on Venus to a maximum of about two hours. Still, Venus has been the subject of very persistent research by space scientists and has yielded enough data about its topography and composition to permit informed speculation about the processes responsible for creating its surface. Scientists' knowledge of the planet's geologic features rests primarily on techniques involving radar imaging, while preliminary impressions of the chemical and structural nature of the surface have been provided through experiments conducted by Soviet spacecraft at several different landing sites.

The Venusian surface is generally smoother than that of any other terrestrial planet. Sixty percent of it lies within 500 meters of Venus's mean radius of 6,051 kilometers. Because Venus has no equivalent to sea level, the mean radius is used as the baseline elevation for topographic measurements. Despite this prevailing uniformity, Venus

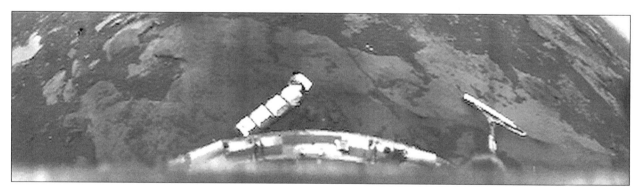

In 1975, the Venera 10 lander survived for sixty-five minutes on the Venusian surface and took this picture of a volcanic-looking surface and part of the spacecraft before it lost function. (NASA)

does have some high mountains and deep valleys. The total range between highest and lowest points on the planet is nearly 14 kilometers, a value that is similar to Earth's.

Planetologists divide Venusian topography into several distinct types of terrain. Rolling plains dominate the globe and form an irregular, planet-girdling area covering more than 70 percent of the surface. About 16 percent of the surface lies below the level of the rolling plains. The remainder is divided among upland plains (0.5-2.0 kilometers above mean radius) and several types of true highlands. Among the latter are the regios, also called domed uplands. These are large, roughly circular areas that rise gently toward their centers, where they achieve heights of between 3 and 5 kilometers above the mean radius. They are thought to be situated over interior "hot spots," which have caused the surface to bubble outward on a gigantic scale. Huge shield volcanoes sit atop many of the domed uplands, a fact that adds credibility to the theory that these landforms are similar to volcanic domes on Earth. Alpha Regio and Beta Regio, the first two surface features identified by Earth-based radar studies, are examples. Surfaces of the domed uplands are generally smooth, like those of the rolling plains, and appear to be the same age. Unlike plains, however, they seem to be criss-crossed by fault lines indicating crustal stresses. Two continent-sized highland areas lie within the mapped region of the surface, but together they account for only 8 percent of the planet's surface. Ishtar Terra and Aphrodite Terra, about the size of Australia and Africa respectively, exhibit a rich variety of landscapes, including two types of mountainous terrain as well as areas of flat and complex plains.

The most common highland topography is one of ridges and valleys that intersect in chevron-shaped or chaotic patterns. This terrain is called tessera terrain, from the Greek word for "mosaic tile," and resembles the deformation patterns that occur on the top of a moving glacier; there are no glaciers on Venus, however. In contrast to the tesserae are more dramatic but less common mountain systems that thrust their peaks 4 to 12 kilometers above the mean radius. The Maxwell Montes

region of Ishtar Terra, which includes the highest known point, 11,800 meters above mean radius, is an example of the latter. It consists of a series of parallel ridges and valleys 15 to 20 kilometers apart. The Maxwell system appears much like the ridge and valley province of the Appalachian Mountains, although it is substantially higher. Its features, and those of at least three other mountain chains on Ishtar Terra, closely resemble those produced when plates of the Earth's crust are thrust together by tectonic forces.

Contrasting with the mountain ranges are great rifts that cleave the surface to depths of up to 2 kilometers. One huge rift system stretches from east to west for more than 20,000 kilometers and can be traced as a series of chasms along the entire southern edge of Aphrodite Terra. From there, it continues across the rolling plains to link with Beta Regio. Another rift splits Beta Regio and continues to Phoebe Regio. This complex system consists of many related but distinct chasms, the largest of which is 3,500 kilometers long and 100 kilometers wide, with its deepest point lying 2.1 kilometers below mean radius.

Among the most interesting surface features thus far discovered are Venus's large volcanoes. Two excellent examples are Thea and Rhea Mons, which, along with

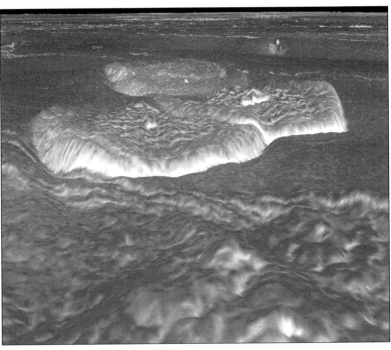

Part of Alpha Regio, a mountainous region with seven domed hills, likely built by outflows of lava. (NASA/JPL)

several other volcanoes, rise from the Beta Regio dome. They appear to be situated on a fault that forms one edge of the great rift. The mass of material that has issued from them is greater than the total output of the volcanic mountains that have formed the Hawaiian Island chain. Both are shield volcanoes like Olympus Mons on Mars, formed by chronic, nonexplosive eruptions of lava that flow long distances before solidifying. Radar images show what may be geologically recent lava flows from both Thea and Rhea, and there is intriguing but very controversial evidence that these volcanoes may have been in eruption in the 1950's and again in the 1970's. Since that time no direct evidence of volcanism has been found by even the long-lived Magellan orbiter.

Impact craters, so characteristic of the surfaces of the Moon and Mercury and even fairly common on Mars, are apparently not nearly so plentiful on Venus. More than one hundred have been observed in radar images, ranging in diameter up to a maximum of 144 kilometers.

Graben—long, linear depressions usually found between parallel faults—are visible in Venus's Themis Regio. These particular graben form a nova, or a series of graben radiating from a central area; about fifty such novae have been found on the Venusian surface. (NASA)

However, the number of craters discovered is considered low, and the largest crater is only modest in size. These facts are considered to be important evidence that the present Venusian surface is not an ancient one. An old surface should bear numerous scars of past encounters with meteors, comets, and asteroids, as is the case with the Moon and Mercury.

Another class of circular geologic feature bearing superficial similarity to impact craters is apparently an unrelated phenomenon. This group comprises the so-called coronae, of which at least eighty have been found. They average 500-800 kilometers in diameter, but their depth is only 200-700 meters, a fact that is inconsistent with an impact origin. Many researchers interpret the coronae to be collapsed bubbles in the crust, caused by localized heating from hot spots in the mantle beneath.

Many Soviet spacecraft failed, but four successfully took panoramic photographs of their landing sites. These pictures are remarkably similar in showing barren landscapes dominated by flat-topped rocks with relatively little loose, fine-grained material that might be described as soil. Closer inspection of images of the rocks shows that they seem to be partially exposed outcroppings of a horizontally layered rock mass that exhibits a marked tendency to break into plate-like slabs. Additional measurements indicate that the rocks are of low density (1.5 grams per cubic centimeter) and high porosity. They have a bearing strength of only a few kilograms per square centimeter, meaning that they can be broken rather easily. These findings are regarded as surprising, for they are characteristic of sedimentary rocks on Earth. Under close inspection, the panoramic photographs seem to support this conclusion, showing what appear to be striations, indicating ripple marks and crossbedding, two common features of sedimentary deposits. Even the electrically nonconductive properties of the rocks, as revealed by their radar reflectivity, agrees well with the behavior of sedimentary rock. If the surface rocks are indeed of sedimentary origin, they presumably formed from deposits of windblown sand.

The velocity of surface winds is low by terrestrial standards, not exceeding 1.3 meters per second. However, under Venus's dense atmosphere, which is ninety times heavier than Earth's, this velocity is more than sufficient to move fine-grained materials and raise dust. At a number of sites on the rolling plains, researchers have detected depressions that seem to be filled with volcanic ash, which may become lithified over time by yet unknown processes.

The mass of the planet, the density of its surface materials, and the relative abundance of certain elements present in its rocks all point to the likelihood that Venus, like Earth, experienced a planet-wide "meltdown" early in its history. The result was differentiation, a process in which the lighter elements migrated to the surface and the heavier elements settled toward the center. Escape of the residual heat from that meltdown is presumed to have been the major architect of the surface features observed on Venus, just as it has been on Earth. Whether the interior remains molten has not been determined. Venus unquestionably lacks a planetary magnetic field; because such a field is thought to be generated by planetary rotation around a molten iron core, this lack suggests that the interior of Venus has cooled and solidified. However, the evidence that volcanism has occurred on the surface in recent geologic time contradicts this view. It may be that Venus, which rotates 243 times more slowly than does Earth, simply does not spin fast enough to create the dynamo effect that gives rise to a magnetic field.

Methods of Study

Study of Venus by optical telescope has not been productive for elucidating the nature of the planet's surface. The first successful attempts to penetrate Venusian clouds were made in 1961. Teams of American, British, and Soviet scientists were able to use large radio telescope antennas to beam radar waves at the planet and receive faint return echoes. Earth-based radar studies of Venus have continued but are seriously limited by the fact that good results can be achieved only when the planet passes near the Earth, at which time Venus always presents the same "face."

To gain a global picture of the Venusian terrain, the National Aeronautics and Space Administration's (NASA's) Pioneer Venus orbiter and the Soviet Venera 15 and 16 spacecraft carried radar-imaging instruments into orbit around Venus. In principle, the American and Soviet spacecraft operated similarly, combining synthetic aperture radar (SAR) imaging with radar altimeter measurements. The SAR images from Pioneer Venus cover 70 percent of the surface but are only about as detailed as those shown by a desktop physical relief globe. Veneras 15 and 16 reached Venus in late 1983. Equipped with larger radar antennas, they were capable of resolving

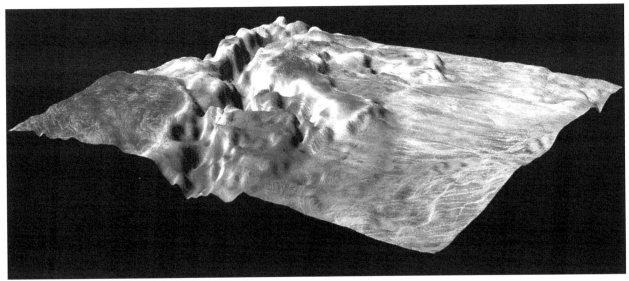

A computer-generated image from Magellan shows a portion of the Aphrodite Terra called Ovda Regio, where plains meet highlands. (NASA/JPL/USGS)

surface details as small as one to two kilometers in size and could measure elevations to within 50 meters. Venera radar images look remarkably like high-altitude black-and-white photographs. Unfortunately, this imaging was obtained only for the northern quarter of the planet (from 90° to 30° north latitude).

Beginning in late 1970, a series of soft landings on Venus were made by Soviet Venera craft equipped to conduct experiments to detect the presence of certain rock-forming minerals. One approach used a gamma-ray spectrometer to detect gamma radiation emitted by radioactive uranium, thorium, and potassium. Venera landers of the 1980's employed an automated drill that bored into the surface to obtain samples not contaminated by chemicals from the atmosphere or from the lander itself. The sample material was transferred to an automated laboratory inside the craft, where it was subjected to X-ray fluorescence. Results showed that the rocks at most of the landing sites appear to resemble basalt, an igneous rock enriched with iron and magnesium. The exact composition, however, varied from site to site and, while not identical to that of Earth basalts, was chemically closer to Earth rocks than to Moon rocks.

At two locations, the experiments detected minerals more characteristic of granite, another common igneous rock. These findings are not seen to be in conflict with evidence that Venusian surface rocks may be sedimentary in nature, for the experiments detected the presence and ratio of identifying minerals but not the type of matrix that contained them.

There is no water on the surface of Venus at present, but two discoveries suggest that such has not always been the case. If water was ever plentiful, it must have boiled away, so that the water molecules were dissociated into oxygen and hydrogen. Investigators have sought evidence for the "missing" oxygen and hydrogen, and some believe that they may have found both. Deuterium, a hydrogen isotope, has been detected to be one hundred times more abundant in the Venusian atmosphere than it is in Earth's. Meanwhile, an experiment has shown that oxidized terrestrial basalts, when heated to the Venusian surface temperature, appear identical in visible and micrometer wavelength imagery to the surface rocks of Venus. The likeliest source for the deuterium and the oxygen to oxidize the basalt is dissociated water molecules.

An intriguing possibility exists that Venus may harbor active volcanoes—perhaps the largest in the solar system (the large Martian volcanoes are definitely extinct). Spacecraft and Earth-based observations have detected large amounts of sulfur dioxide, a common volcanic effluent, in the Venusian atmosphere. Moreover, sulfur dioxide content increased dramatically in the 1950's and again in the 1970's.

The first color pictures taken on the surface by the Venera 13 lander seemed to show that the landscape had an orange or amber tint, which proved to be an effect of sunlight filtered through the heavy overcast. Computer processing of the photographs has since shown that, in normal white light, the rocks are a uniform, colorless gray.

Previous spacecraft had performed preliminary radar investigations of Venus's surface, identifying the major types of features on that surface and identifying prominent examples of each. However, high-resolution maps of the entire surface were lacking. The goal of NASA's Magellan spacecraft was to use a synthetic aperture radar in prolonged orbit about Venus to produce a global map at a resolution even in excess of the best contemporary maps of Earth's surface. Detailed geological interpretations and altimetry data were obtained in the process.

Magellan launched aboard the space shuttle Atlantis on May 4, 1989, and was the primary payload of the STS-30 mission. The spacecraft was deployed from the shuttle's cargo bay, and was dispatched on a trajectory that concluded with orbital insertion about Venus on August 10, 1990. Magellan entered a highly elliptical orbit, often ranging from as little as 300 to as much as 8,500 kilometers above the surface. With Magellan in a polar orbit, the planet Venus rotated underneath it, allowing the spacecraft to image a different ground track on each low pass. The spacecraft turned toward Earth as it climbed toward its highest orbital point and then transmitted the radar imagery it had collected during its low pass over Venus.

Science activities were slightly varied with each of Magellan's six cycles, with the spacecraft's orbit occasionally being altered for different research requirements. During the first cycle, Magellan concentrated on global radar mapping and imaged 84 percent of Venus. Later cycles filled in gaps and concentrated on specific features of interest. In a lower orbital altitude late in its operational mission, Magellan was able to collect precise gravitational data as Venus slightly altered the spacecraft's orbital parameters. Magellan was used to test aerobraking techniques by having the spacecraft fly through the upper portions of Venus's atmosphere; its large solar panels experienced a retarding torque due to atmospheric drag. How the spacecraft responded to the atmosphere indirectly informed scientists about Venus's atmospheric particle density as a function of altitude. With its primary and extended missions completed, flight controllers decided to send Magellan plunging into the upper atmosphere to remove it from orbit. Maneuvers were conducted to force the spacecraft's orbit to decay due to orbital drag. On October 11, 1994, the final spacecraft maneuver was conducted. Controllers lost contact with Magellan the following day. Then, on October 14, Magellan was destroyed in the atmosphere. Although it could not be verified, many believed pieces of descending debris survived long enough to impact the surface.

Perhaps the most important results of Magellan's intense investigation of Venus was determining a total lack of plate tectonics based on the two primary processes observed on Earth. Instead of continental drift and basin floor spreading, Venus's global rift zones and coronae move as a result of upwelling and subsidence of magma in the planet's mantle. That suggested that Venus's surface is indeed quite young geologically speaking, perhaps less than 800 million years old. The enormous data set from Magellan was made available to interested researchers and individuals on compact disc.

Despite the extensive research with Magellan, many questions remained to be investigated. The European Space Agency (ESA) dispatched its Venus Express spacecraft to Venus in order to examine the planet's atmosphere at infrared wavelengths. Venus Express launched on November 9, 2005, and was inserted successfully into Venus orbit on April 11, 2006. Its Visible and Infrared Thermal Imaging Spectrometer (VIRTIS) was used to identify the amount of sulfur dioxide in the atmosphere between 35 and 40 kilometers and to monitor that constituent for changes in concentration over time that would indicate active volcanism. Venus Express's Spectroscope for Investigation of Characteristics of the Atmosphere of Venus (SPICAV) used stellar occultation methods to determine the identity of atoms and molecules in the upper atmosphere at an altitude between 70 and 90 kilometers. SPICAV saw rapid drops in the amount of sulfur dioxide in the upper atmosphere, strongly indicating that Venus has active volcanoes. VIRTIS was then used to identify hot spots on the surface.

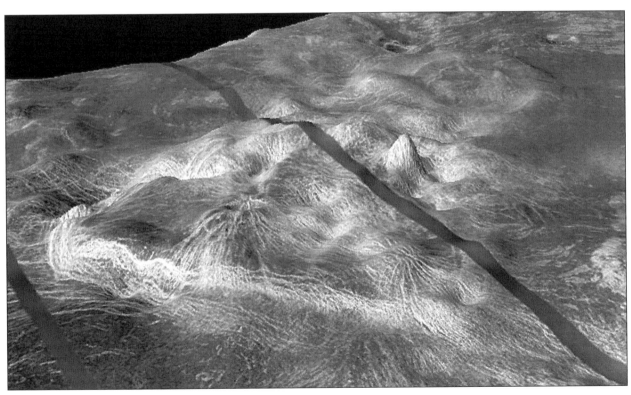

Yavine Corona, one of many coronae, circular regions averaging 500-800 kilometers in diameter, possibly collapsed crustal bubbles caused by localized heating from hot spots in the mantle beneath. Yavine Corona contains two novae. (NASA/JPL/USGS)

Context

In order to understand a system as complex as a terrestrial planet, it is necessary to have more than one example of how such planets function and evolve. Hence, the study of geologic processes on another world, far from being simply an esoteric and impractical inquiry, holds promise for improving human understanding of the forces that have acted on Earth and still continue to shape its surface. For this reason, the goal of Venusian geological studies is to discover the relationships and sequences of events that have resulted in the landforms that can be imaged. "Looking at Venus is like running the experiment that produced the Earth a second time," according to Robert Kunzig, senior editor of *Discover* magazine. Indeed, scientists appreciate the opportunity to "run the experiment" again under slightly different conditions in order to see whether their set of explanatory theories can accommodate any observed deviations in the results. Venus presents a marvelous opportunity to test the plate tectonics theory on a planet that has many fundamental similarities to Earth but also exhibits numerous significant differences.

It is generally believed that Earth's loss of internal heat is primarily a result of convection and occurs mainly along the 75,000-kilometer-long mid-ocean ridge. Seafloor spreading that results is responsible for producing a large expanse of young and renewable crust. Even the older continental masses are invigorated by the tectonic activity that is driven by this convective heat loss. Venus seems to have experienced similar horizontal crustal movements in the past and may still be experiencing them. The driving force, however, might be quite different. Most authorities interpret the present surface as having been formed through the release of heat at localized hot spots.

Another fundamental question is whether rocks making up Venus's vast, rolling plains differ significantly in composition from those of the higher terrain. This issue is of interest because it bears on where and how crustal materials originated. It is not inconceivable that Venus could reveal hitherto unknown relationships that may have acted on Earth when the original continental rocks were first solidifying some 3.8 billion years ago.

The mystery of whether Venus once had oceans is of particular interest for two reasons. First, most scientists now accept as true evidence that Earth's own atmosphere is beginning to warm as a result of increases in carbon dioxide content. There is growing concern that atmospheric pollution may trigger a runaway "greenhouse effect," similar to the process that appears to have happened on Venus. If Venus retains any "memory" of conditions before it became so hot, it will only be in the record of the rocks themselves. Second, scientists are still uncertain about how Earth got its abundant water in the first place. If Venus had oceans at a former time, that fact would have significant implications for theories of the origin of planetary water.

A large amount of the Venusian surface was mapped by Pioneer Venus's radar, but its images were good enough only to show gross features of the surface and could not address cause-and-effect relationships. Better imaging has been obtained by the Arecibo radio telescope and the Soviet Venera 15 and 16 spacecraft, but the total area covered was too small to permit generalizations to be drawn. Together, all these data allowed planetary scientists the luxury of asking better questions, which might then be answered by the higher-resolution Magellan spacecraft's synthetic aperture radar imaging system. Magellan produced a spectacular increase in knowledge of Venusian topography. Magellan imagery provided planetary scientists with the information needed to correlate the roles of volcanic activity, tectonic motion, and impact events in the formation and evolution of Venus's surface features. Magellan established that some surface features resulted from tectonics and Venusian volcanoes have been active in recent geologic time, but some key questions are likely to remain unanswered until more complicated surface experiments can be conducted.

Richard S. Knapp

Further Reading

Bazilevskiy, Aleksandr T. "The Planet Next Door." *Sky and Telescope* 77 (April, 1989): 360-368. The author, a senior member of the Soviet Venera science team, provides a comprehensive overview of the surface of Venus in clear and nontechnical terms. The article is particularly valuable for its discussion of Venus's medium- and small-scale surface features and for its fair-minded discussion of issues about which there is significant debate or uncertainty.

Bredeson, Carmen. *NASA Planetary Spacecraft: Galileo, Magellan, Pathfinder, and Voyager.* New York: Enslow, 2000. This book, part of Enslow's Countdown to Space series, provides an overview of NASA planetary exploration during the last two decades of the twentieth century. Suitable for all audiences.

Burgess, Eric. *Venus: An Errant Twin.* New York: Columbia University Press, 1985. Probably the general reader's most complete single source of information about Venus and how present knowledge has been obtained. Chapters on the Veneras, Pioneer Venus, and the relationship of Venus's geological history to that of Earth and Mars round out the detailed discussion of the surface and atmosphere. Nearly one hundred well-chosen photographs and diagrams illustrate the text.

Cattermole, Peter John. *Venus: The Geological Story*. Baltimore: Johns Hopkins University Press, 1996. Provides a comprehensive presentation of the latest understanding of Venus, based on Magellan data.

De Pater, Imke, and Jack J. Lissauer. *Planetary Sciences*. New York: Cambridge University Press, 2001. A challenging and thorough text for students of planetary geology, this volume offers an excellent reference for the most serious reader with a strong science background. Provides an in-depth contemporary explanation of solar-system formation and evolution.

Faure, Gunter, and Teresa M. Mensing. *Introduction to Planetary Science: The Geological Perspective*. New York: Springer, 2007. Designed for college students majoring in Earth sciences, this textbook provides an application of general principles and subject material to bodies throughout the solar system. Excellent on comparative planetology.

Fimmel, Richard O., Lawrence Colin, and Eric Burgess. *Pioneering Venus: A Planet Unveiled*. Washington, D.C.: National Aeronautics and Space Administration, 1995. A profusely illustrated scientific and technical publication from NASA that includes the Pioneer Venus data as well as a good deal of information from the Russian spacecraft dispatched to investigate Venus.

Grinspoon, David Harry. *Venus Revealed: A New Look Below the Clouds of Our Mysterious Twin Planet*. New York: Basic Books, 1998. Incorporates Magellan mapping and other data in its coverage of Venusian geology; explains the Venusian greenhouse effect; speculates about Venus's past. A must for the planetary science enthusiast who wants an integrated approach to science and history.

Harvey, Brian. *Russian Planetary Exploration: History, Development, Legacy, and Prospects*. New York: Springer, 2007. Early Russian space programs attempted a large number of Moon, Venus, and Mars investigations. Many were successful, many not. These robust programs are often overlooked. This is their story in one illuminating book about the engineering, development, flight operations, and science returns.

Kerr, Richard. "Venusian Geology Coming into Focus." *Science* 224 (May 18, 1984): 702-703. A brief, easily under-stood summary of the major geologic features revealed by radar mapping. Interpretation of the Venera 15 and 16 results by leading American planetologists forms the basis of the article. *Science* frequently publishes articles dealing with research on Venus. Many are suitable for lay readers and those with a general science background.

Marov, Mikhail Ya, and David Grinspoon. *The Planet Venus*. New Haven, Conn.: Yale University Press, 1998. Marov was Soviet Venera mission chief scientist, Grinspoon a NASA-funded scientist studying Venus. Together they provide a coordinated description of American and Soviet attempts to learn the secrets of Venus, a planet shrouded in mystery. For both general readers and specialists.

Morrison, David, and Tobias Owen. *The Planetary System*. 3d ed. San Francisco: Pearson/Addison-Wesley, 2003. A full chapter is devoted to the geologic and atmospheric processes on Venus. An additional chapter covers the origin of the solar system, useful to those not familiar with current theories on planetary formation. Each of the other terrestrial bodies also receives a chapter of discussion.

Spanenburg, Ray, and Kit Moser. *A Look at Venus*. New York: Franklin Watts, 2002. A look beneath the thick clouds of Venus. Written for a younger audience.

Young, Carolynn, ed. *The Magellan Venus Explorers' Guide*. Pasadena, Calif.: Jet Propulsion Laboratory, California Institute of Technology, National Aeronautics and Space Administration, 1990. Prepared as a field and educational guide to the Magellan mission. Published prior to mission launch, this volume contains no spacecraft results, but it does describe the expected research and its value to Venus studies.

VENUS'S VOLCANOES

Categories: Planets and Planetology; Venus

The planet Venus has at least sixteen hundred major volcanoes and many more minor ones, which is more volcanoes than any other planet in the solar system. Most are shield volcanoes, but Venus also has pancake domes and other volcanic features. About 80 percent of Venus's surface has been shaped by some type of volcanic activity. Venus does not have volcanic chains like those formed on Earth from plate tectonics. Comparing volcanic features on Venus and Earth helps us better understand volcanic processes on both planets.

Overview

Venus is sometimes referred to as Earth's twin sister because its size, mass, and density are similar to those of Earth. Venus's surface conditions, however, definitely make Venus Earth's "evil" twin sister. Owing to a runaway greenhouse effect from the carbon dioxide atmosphere, Venus has a surface temperature hot enough to melt lead. The surface atmospheric pressure is nearly one hundred times what it is on Earth. Thick layers of sulfuric acid clouds veil the surface of Venus. Landers on Mars can last for years, but on Venus they are destroyed by the harsh surface conditions within about an hour. These atmospheric conditions make it impossible to study the surface of Venus using direct optical means or long-term robotic landers. Astronomers must use radar maps rather than optical photographs to study the planet's surface features. Radar maps from both Earth and spacecraft have, however, unveiled the surface of our mysterious twin sister.

Radar maps show that volcanic activity has played a major role in shaping the surface of Venus. Volcanic activity includes not only erupting volcanoes but also lava flows and other activity whereby solid, liquid, or gaseous material escapes from the planet's interior. Volcanic activity is often caused by tectonic activity but can occur independently of tectonic activity. There are more than sixteen hundred large volcanic features on the surface of Venus and possibly as many as hundreds of thousands of smaller volcanic features. In addition, about 80 percent of the planet's surface is covered with flat plains that are probably solidified lava flows. These lava plains formed when lava flooded areas covering thousands of square kilometers and then solidified.

Most of the volcanoes on Venus are shield volcanoes. Shield volcanoes derive their name from their resemblance to ancient warriors' shields lying on the ground pointing upward. Shield volcanoes are often very large, but they have fairly gentle, rather than very steep, slopes. Shield volcanoes form when lava flows out from a single central vent. Rather than forming on the boundaries of tectonic plates, shield volcanoes usually form over a volcanic hotspot. These hotspots are places in the planet's crust where lava wells up from the planet's mantle. When the lava breaks through the surface, it erupts to form a shield volcano. Successive eruptions can form very large shield volcanoes.

Two of the larger known shield volcanoes on Venus are Sif Mons and Gula Mons. They have peak altitudes of about 4 kilometers above the surrounding surface, which compares to Mauna Loa, Earth's largest shield volcano, which rises about 8 to 9 kilometers above the Pacific ocean floor. (The largest shield volcano in the solar system is Mars's Olympus Mons, which towers about 25 kilometers above the Martian surface.) Often the top of a shield volcano will collapse to form a crater, known as a caldera. This collapse occurs when the lava flow retreats back to the planet's mantle, leaving nothing to support the top of the volcano. The calderas formed on Sif Mons and Gula Mons are about 100 kilometers across. Calderas are fairly common on the surfaces of both Venus and Earth. Calderas, however, are not the only types of craters found on Venus. Venus has many large impact craters that formed from meteorite impacts rather than volcanic activity.

The largest volcanic features found on Venus are coronae. These features are not found on the other

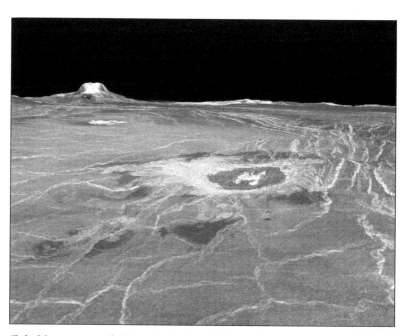

Gula Mons, seen on the Venusian horizon (left), rises to an altitude of about 3 kilometers and is one of the larger shield volcanoes on Venus. Cunitz Crater can be seen in the center middleground. This image was returned by the Magellan spacecraft. (NASA/JPL)

terrestrial planets. Coronae, which are approximately circular in shape (hence their name, from the Latin for "crown"), form from an uplifting process. Hot mantle material swells and pushes the crust upward. Coronae usually have associated volcanoes and lava flows. Aine is a large corona on Venus that is about 300 kilometers in diameter. On a larger scale, Lakshmi Planum, which is part of Ishtar Terra—one of Venus's two large continental sized features—is about 1,500 kilometers at its widest point. Lakshmi Planum likely formed from the same process that formed the coronae, but on a larger scale.

Another common type of volcano found on Venus is the lava dome or pancake dome. Venus's lava domes are much smaller than its shield volcanoes, being typically tens of kilometers in diameter or less. They are usually circular and relatively flat—hence the name "pancake dome." They form when lava slowly flows out onto the surface and then flows back. However, a thin crust solidifies on the surface of the lava. When the lava subsides, the crust stays and cracks because it lacks support. Pancake domes are often found near coronae.

On Earth, volcanoes often form at the boundaries of the tectonic plates. Examples are the volcanoes on the western coasts of North and South America and the Mid-Atlantic Ridge, which runs along the Atlantic Ocean's floor between the North American and Eurasian plates. Such volcanoes are not found on Venus. Venus apparently does not have tectonic plates on its crust. Venus does not have plate tectonics similar to Earth's, but it does have tectonic activity. On Earth, plate tectonics is caused by convection currents in Earth's mantle slowly moving the crustal plates horizontally. On Venus, the crust is not divided into plates. Convection currents in the mantle cause vertical rather than horizontal crustal movement on Venus. Coronae are a good example of volcanic features formed on Venus from the crust's vertical tectonic motion.

Are the volcanoes on Venus still active, as on Earth, or are they extinct, as on Mars? Planetary scientists do not yet know the answer to this question. Despite being volcanically active, at any given time few of Earth's many volcanoes are actively erupting. The same would be true on Venus. Hence scientists would not expect to see volcanoes continually erupting on Venus, even if it is still volcanically active. Partly because of the planet's thick cloud layer, no one has observed a volcanic eruption on Venus, but there is some indirect circumstantial evidence to suggest that volcanoes on Venus are still active. Volcanic eruptions emit sulfur dioxide gas. Scientists observe frequent variations in the amount of sulfur dioxide in Venus's upper atmosphere. These variations could be caused by occasional volcanic eruptions spitting sulfur dioxide into the atmosphere. Space probes to Venus have also detected radio outbursts from Venus that are similar to those produced by lightning discharges from erupting volcanoes on Earth. These observations are evidence, but not proof, that Venus's volcanoes are still active. If planetary scientists were to observe a volcano on Venus in the act of erupting, then Venus would join Earth and Jupiter's satellite Io as the worlds in the solar system with still-active volcanoes.

A close-up computer simulation of Gula Mons was generated from data returned by the Magellan spacecraft. It rises 3 kilometers above the area known as Eistla Regio. (NASA/JPL)

Knowledge Gained

Because thick clouds veil the surface of Venus, astronomers for a long time could only speculate about the planet's surface. Prior to the space age, speculations varied: The surface was envisioned as hot and steamy by some, as a hot and dry desert by others. With the coming of the space age, astronomers were finally able to gather data on the surface of Venus. They did not, however, suspect just how hot Venus really was.

The first radar maps of Venus from Earth were made using the Arecibo radio telescope in Puerto Rico, beginning in the late 1970's. Because of the distance to the planet, these images had a relatively low resolution, on the order of a few kilometers. These Earth-based radar maps did, however, allow planetary scientists to observe the large-scale surface features of Venus.

Earth-based radar maps of Venus can reveal only part of the Venusian surface, because during Venus's closest approach to Earth only one side faces Earth. Orbiting spacecraft have therefore been sent to map the surface of Venus. Pioneer Venus 1 went into orbit around Venus in late 1978. This mission was the first orbital mission to use radar to map much of the surface of Venus, with a resolution of about 7 kilometers. In 1990, the Magellan mission went into orbit around Venus. The Magellan orbiter made much more extensive and detailed radar maps of Venus. Because it used a polar rather than an equatorial orbit, Magellan was able to map essentially the entire surface, including the polar regions, which were hidden to previous missions. The best resolution of the Magellan radar maps is about 100 meters.

The fact that Venus has volcanic activity, including both volcanic mountains and lava plains, indicates that, geologically speaking, the surface of Venus is very young. The surface, not the planet itself, is probably less than a half a billion years old.

Context

The volcanoes on Venus contribute to its very harsh surface environment. Volcanoes on Earth outgas significant amounts of carbon dioxide gas. Those on Venus and Mars probably do the same. On Earth, biological activity, such as plant respiration, uses the carbon dioxide. Earth therefore has a very small percentage of carbon dioxide in its atmosphere. On Venus, however, the carbon dioxide is still in the atmosphere; 97 percent of Venus's atmosphere is carbon dioxide. All this carbon dioxide produces a runaway greenhouse effect and surface temperatures greater than 700 kelvins.

Jupiter's moon Io is also volcanically active. However the volcanoes on Io differ from the volcanoes on Venus and other terrestrial planets. In addition to rock, Io's composition includes significant amounts of ice.

Venus, Mars, and Earth all have large volcanoes. However, the volcanic and tectonic activity is different on each of these three planets. The differences arise from differences in size and internal heating of the planets. Mars has had the least amount of volcanic activity. Earth's crust is divided into several tectonic plates. Movement of these plates is an important force in shaping the volcanic activity on Earth. Venus does not have a crust broken into several plates. Hence, Venus has much volcanic activity, but it does not have the types of features formed by crustal plate movement. Understanding how volcanic and tectonic activity differs among the various planets is one of the frontiers of planetary science.

Paul A. Heckert

Further Reading

Chaisson, Eric, and Steve McMillan. *Astronomy Today.* 6th ed. New York: Addison-Wesley, 2008. Chapter 9 of this very readable introductory astronomy textbook covers the planet Venus. There is a good section on the planet's surface volcanism.

Freedman, Roger A., and William J. Kaufmann III. *Universe.* 8th ed. New York: W. H. Freeman, 2008. Chapter 11 of this introductory astronomy textbook is a complete and readable overview of Mercury, Venus, and Mars, including volcanic activity.

Hartmann, William K. *Moons and Planets.* 5th ed. Belmont, Calif.: Thomson Brooks/Cole, 2005. This textbook on the planets and satellites of the solar system summarizes our understanding of volcanic and other tectonic processes on Venus as well as other planets and moons.

Hester, Jeff, et al. *Twenty-first Century Astronomy.* New York: W. W. Norton, 2007. Chapters 6 and 7 of this well-illustrated astronomy textbook are about the terrestrial planets. Volcanic and tectonic processes are well covered, and the comparison of these processes on different planets is very good.

Morrison, David, Sidney Wolf, and Andrew Fraknoi. *Abell's Exploration of the Universe.* 7th ed. Philadelphia: Saunders College Publishing, 1995. Venus is covered in chapter 15 of this classic astronomy textbook.

Zeilik, Michael. *Astronomy: The Evolving Universe.* 9th ed. New York: Cambridge University Press, 2002. Chapter 9 of this astronomy textbook provides an overview of the terrestrial planets.

Zeilik, Michael, and Stephen A. Gregory. *Introductory Astronomy and Astrophysics.* 4th ed. Fort Worth, Tex.: Saunders College Publishing, 1998. Pitched at undergraduate physics or astronomy majors, this more advanced textbook goes into greater mathematical depth than most introductory astronomy texts. Chapters 4 and 5 cover the basic principles of the terrestrial planets, including volcanic processes.

VIKING PROGRAM

Category: Space Exploration and Flight

The Viking Program sent two uncrewed craft, Viking 1 and Viking 2, to Mars. These probes provided the first data from the surface of Mars and detailed images of the entire planet. The two landers were the first craft to land on and directly study the surface of Mars, greatly increasing the knowledge of Mars.

Overview

Prior to the establishment of the Viking Program, four Mariner missions had been sent to Mars. The first, Mariner 4, was launched in 1964. It returned twenty-one images as it flew to within 10,000 kilometers of Mars. Five years later, Mariner 6 and Mariner 7 flew past Mars and sent back 201 pictures. Mariner 9 reached Mars in 1971. Because it was an orbiter rather than a flyby, Mariner 9 returned 7,300 images during its one-year lifetime. This mission discovered the volcanoes and Valles Marineris, which were explored in more detail by the Viking Program. In addition to the American Mariner probes, the Soviet Union launched four largely unsuccessful probes to Mars in 1973, including one that attempted to land on the surface.

Development of the Viking Program

Although plans to conduct uncrewed missions to Mars were initiated shortly after the National Aeronautics and Space Administration (NASA) was established in 1958, the Viking Program was not approved until 1968. Its primary scientific goal was to search for evidence of life on Mars. Other goals were to land on the surface of Mars and return scientific data about the planet. After the project's original launch date was delayed from 1973, the Vikings were finally launched in 1975 from Cape Canaveral, Florida, atop Titan III launch vehicles. They spent one year traveling to Mars and arrived in the summer of 1976.

Each Viking craft consisted of both an orbiter and a lander. Both Viking orbiters orbited the planet for a few years before being powered down. After 706 orbits, the Viking 2 orbiter was turned off first on July 25, 1978. The Viking 1 orbiter completed more than 1,400 orbits before being turned off on August 17, 1980. The two orbiters had different orbital inclinations, so that Viking 1 could study the lower latitudes, while Viking 2 could study the polar regions. Both orbiters returned what at the time were the highest-resolution images available of the entire Martian surface. They also made various atmospheric measurements. Each of the orbiters also flew near one of the two moons of Mars, returning close-up images.

Shortly after arrival at Mars, each lander separated from its orbiter to land on the surface of Mars. Viking 1 landed on the Chryse Planitia, or plain, at 22 degrees north latitude and 47 degrees west longitude on July 20, 1976. Viking 2 landed on the Utopia Planitia at 48 degrees north latitude and 226 degrees west longitude on September 3, 1976. Plains or relatively flat areas were chosen for both landing sites in order to minimize the risk of a crash landing. Despite this strategy, Viking 1 missed a large boulder by only 7.5 meters (25 feet). These two Viking landers were the first craft to successfully land on the surface of Mars. The Viking 2 lander continued to return data until the end of its mission on April 11, 1980. The Viking 1 lander continued to return data until the end of its mission on November 13, 1982. The data included detailed images of the surface, analysis of the soil and atmosphere, and a negative search for evidence of life.

Viking Experiments

Each Viking orbiter contained two cameras used to obtain detailed images of the entire Martian surface. The two Viking orbiters took a combined total of more than 46,000 photographs of the Martian surface. The average resolution was from 150 to 300 meters. In March, 1977, the Viking 1 orbit was adjusted so that its closest approach to Mars was 300 kilometers. Viking 2 followed on October 23, 1977. At this distance, the Viking orbiters could resolve surface features as small as 20 meters, although they did not map the entire surface with this resolution.

After performing their initial task of locating safe landing sites for the landers, the orbiter cameras revealed details of both previously known and previously undiscovered features, including global dust storms;

the Olympus Mons and Tharsis Ridge volcanoes, the largest known volcanic mountains in the solar system; the Valles Marineris, a canyon that would stretch across the entire United States; and a number of arroyos, or dry river beds, and other features indicating the past presence of large amounts of liquid water. The orbiters also contained infrared spectrometers and radiometers to map Mars's atmospheric water vapor content and thermal properties. The Viking 1 orbiter took pictures of Phobos, the larger of Mars's two moons, from a closest approach of 90 kilometers. The Viking 2 orbiter photographed Deimos, the smaller of Mars's two moons, from a closest approach of 22 kilometers.

During the landers' descents, instruments on the landers performed analysis of the Martian atmospheric properties at various levels to determine the atmospheric structure. These properties included composition, pressure, and temperature. Scientists learned that the atmosphere of Mars is primarily (95 percent) carbon dioxide. At about 0.5 percent of Earth's surface atmospheric pressure, the surface pressure of Mars is too low for water to exist in a liquid state. Coupled with the evidence of large amounts of liquid water in the past, this information provided evidence for major global climate changes on Mars. After landing, these atmospheric instruments continued to provide data on the Martian weather and climate.

Each Viking lander contained two cameras to provide direct close-up images of the Martian surface. Over the course of the Viking mission, a total of more than 1,400 images were sent back from the Martian surface. They revealed a boulder-strewn reddish surface, which gives Mars the red color that inspired the ancients to name the planet for their god of war. Because Mars's atmosphere is too thin to scatter blue light the way Earth's atmosphere does and because the airborne dust particles are red in color, the images also show a pink sky. The images also show early morning surface frost and structures like sand dunes that result from the global dust storms.

To analyze the composition of this red surface, each lander was equipped with a scoop that could dig a few centimeters into the surface and place a surface sample into an X-ray fluorescence spectrometer. Analysis of these samples showed that the Martian surface contains a fairly large percentage of both iron and oxygen. The red color is iron oxide or rust.

Each lander also contained a seismometer to measure earthquake, or marsquake, activity. The Viking 1 seismometer failed, the only instrument to fail in the entire

Viking mission. The other seismometer revealed that the few quakes that occurred were very weak.

Scientists integrated the results of these various experiments to deduce the geological and atmospheric history of Mars. They determined that whereas Mars had been geologically active early in its history, that geologic activity had ceased. Although the planet's volcanoes were no longer active, they had at one time released carbon dioxide, water vapor, and other gases. During this time, the thicker atmosphere had allowed water to exist in liquid form. As the atmosphere gradually escaped into space, water could no longer exist as a liquid and is now present as ice and vapor in thin clouds. Carbon dioxide is still present in the atmosphere. Solar ultraviolet rays split much of the water into hydrogen and oxygen. The lighter hydrogen escaped into space. The oxygen combined with the iron on the surface to produce the rust and the red color.

The Search for Life on Mars

Perhaps the most important experiments performed by the Viking landers were the three experiments designed to search for signs of life on Mars. Although cameras had already observed no obvious signs of large life forms, these experiments—the gas-exchange, labeled-release, and pyrolytic-release experiments—were designed to look for evidence of microscopic life.

The gas-exchange experiment involved placing a soil sample into an aqueous nutrient solution that was dubbed "chicken soup" by the experimenters. Any primitive life forms using the nutrients should release gas, which could be detected by looking for changes in the atmospheric composition of the test chamber. This experiment found no evidence of biological activity.

The labeled-release experiment used carbon-14, a radioactive isotope of carbon, in the nutrient solution. Primitive microscopic organisms absorbing the nutrient would eventually release gas containing the carbon-14, which could be detected by its radioactivity. The pyrolitic-release experiment also used carbon-14, but it used it in the atmosphere rather than in the nutrients. After an incubation period, the sample was analyzed to see if it had absorbed any of the carbon-14. The results of these experiments were somewhat ambiguous, but could be explained by chemical rather than biological reactions. Hence, the Viking mission, which was the first specific attempt to find evidence of life on another planet, found no solid evidence of life on Mars.

Prior to the inception of the U.S. space program, many people had believed that Mars might contain life. This idea resulted largely, but not entirely, from the efforts of Percival Lowell who, in the early twentieth century, had popularized the idea that Mars contained canals. Lowell made detailed maps of these alleged canals, asserting that they had been built by a race of Martians to transport water from the pole caps to the warmer equatorial regions on their dry world. Although most astronomers of the time disputed Lowell's work, the idea of life on Mars had been firmly planted in the public imagination. In addition, other scientific studies of Mars had shown that although conditions there might be harsh, life on the planet was at least a possibility. After the Viking Program found no evidence of life on Mars, most scientists assumed that Mars was lifeless. However, the question was reopened two decades later by the announcement of possible evidence for primitive fossilized bacteria in a Martian meteorite that had been discovered in Antarctica.

After the Viking Program ended, there was relatively little exploration of Mars until the 1990's, when exploration of Mars resumed. The next successful landing on Mars was that of the Pathfinder, which landed in July, 1996, twenty years after the Vikings had landed. For two decades, the Viking Program had provided humankind's most detailed knowledge of the planet Mars.

Paul A. Heckert

Further Reading

Hartmann, William K. *Moons and Planets*. 3d ed. Belmont, Calif.: Wadsworth, 1993. Written from a comparative planetology perspective, this book contains information on Mars integrated throughout the text and organized by specific topics, such as atmospheres and interiors. Written between the Viking and more recent Mars missions, the information on Mars is primarily that of the Viking mission.

Moore, Patrick, and Garry Hunt. *Atlas of the Solar System*. Chicago: Rand McNally, 1983. The chapter on Mars contains a brief history of Martian observations prior to Viking, a summary of the Viking mission, and a summary of the mission's results. The book also contains a large number of photographs and maps of the Martian surface.

Morrison, David, and Tobias Owen. *The Planetary System*. Reading, Mass.: Addison-Wesley, 1988. Chapters 9 and 10, on Mars, concentrate on the scientific knowledge gained from the Viking mission.

Raeburn, Paul. *Mars: Uncovering the Secrets of the Red Planet*. Washington, D.C.: National Geographic Society, 1998. Filled with high-quality photographs, this book tells the story of Mars exploration. Chapter 3 concentrates on the Viking mission.

INDEX